Let's Keep in Touch

Follow Us   Online

Visit US at

www.EffortlessMath.com

 https://www.facebook.com/Effortlessmath

Call

 https://goo.gl/2B6qWW

1-469-230-3605

Online Math Lessons

It's easy! Here's how it works.

1- Request a FREE introductory session.

2- Meet a Math tutor online via Skype.

3- Start Learning Math in Minutes.

Send Email to: info@EffortlessMath.com

Or Call: **+1-469-230-3605**

www.EffortlessMath.com

... So Much More Online!

- **FREE Math lessons**

- **More Math learning books!**

- **Online Math Tutors**

Looking for an Online Math Tutor?

Call us at: 001-469-230-3605

Send email to: Info@EffortlessMath.com

Arithmetic and Pre-Algebra Workbook

Comprehensive Activities for Mastering Essential Math Skills

By

Reza Nazari

& Ava Ross

Arithmetic and Pre-Algebra Workbook

Copyright © 2017

Reza Nazari & Ava Ross

All rights reserved. No part of this publication may be reproduced, stored in a retrieval system, or transmitted in any form or by any means, electronic, mechanical, photocopying, recording, scanning, or otherwise, except as permitted under Section 107 or 108 of the 1976 United States Copyright Ac, without permission of the author.

All inquiries should be addressed to:

info@effortlessMath.com

www.EffortlessMath.com

ISBN-13: 978-1981441891

ISBN-10: 1981441891

Published by: Effortless Math Education

www.EffortlessMath.com

Description

Arithmetic and Pre-Algebra Workbook provides students with the confidence and math skills they need to succeed in any math course they choose and prepare them for future study of Algebra, Geometry, Pre-Calculus and Calculus, providing a solid foundation of basic Math topics with abundant exercises for each topic. It is designed to address the needs of math students who must have a working knowledge of basic Math and pre-algebra.

Inside the pages of this comprehensive Workbook, students can learn basic math operations in a structured manner with a complete study program to help them understand essential math skills. It also has many exciting features, including:

- Dynamic design and easy-to-follow activities
- A fun, interactive and concrete learning process
- Targeted, skill-building practices
- Fun exercises that build confidence
- All solutions for the exercises are included, so you will always find the answers

Arithmetic and Pre-Algebra Workbook is an incredibly useful tool for those who want to review all topics being taught in Pre-algebra courses. It efficiently and effectively reinforces learning outcomes through engaging questions and repeated practice, helping you to quickly master basic Math skills.

About the Author

Reza Nazari is the author of more than 100 Math learning books including:
– **Math and Critical Thinking Challenges:** For the Middle and High School Student
– **GED Math in 30 Days.**
– **ASVAB Math Workbook 2018 - 2019**
– **Effortless Math Education Workbooks**
– and many more Mathematics books ...

Reza is also an experienced Math instructor and a test–prep expert who has been tutoring students since 2008. Reza is the founder of Effortless Math Education, a tutoring company that has helped many students raise their standardized test scores—and attend the colleges of their dreams. Reza provides an individualized custom learning plan and the personalized attention that makes a difference in how students view math.

To ask questions about Math, you can contact Reza via email at:
reza@EffortlessMath.com

Find Reza's professional profile at:
goo.gl/zoC9rJ

Arithmetic and Pre-Algebra Workbook

Contents

Description 2

Chapter 1: Arithmetic 10

1-1 Simplifying Fractions 11

1-2 Adding and Subtracting Fractions 12

1-3 Multiplying and Dividing Fractions 13

1-4 Adding and Subtracting Mixed Numbers 14

1-5 Multiplying and Dividing Mixed Numbers 15

1-6 Adding and Subtracting Decimals 16

1-7 Multiplying and Dividing Decimals 17

1-8 Converting Between Fractions, Decimals and Mixed Numbers 18

1-9 Rounding Numbers 19

1-10 Factoring Numbers 20

1-11 Greatest Common Factor 21

1-12 Least Common Multiple 22

Answers of Worksheets – Chapter 1 23

Chapter 2: Real Numbers and Integers 28

2-1 Adding and Subtracting Integers 29

2-2 Multiplying and Dividing Integers 30

2-3 Ordering Integers and Numbers 31

2-4 Arrange, Order, and Comparing Integers 32

2-5 Order of Operations 33

2-6 Mixed Integer Computations 34

2-7 Absolute Value 35

2-8 Integers and Absolute Value 36

2-9 Classifying Real Numbers Venn Diagram 37

Answers of Worksheets – Chapter 2 38

Arithmetic and Pre-Algebra Workbook

Chapter 3: Proportions and Ratios 41
3-1 Writing Ratios 42
3-2 Simplifying Ratios 43
3-3 Proportional Ratios 44
3-4 Create a Proportion 45
3-5 Similar Figures 46
3-6 Similar Figure Word Problems 47
3-7 Simple Interest 48
3-8 Complete the Ratio Table 49
3-9 Ratio and Rates Word Problems 50
Answers of Worksheets – Chapter 3 51
Chapter 4: Percent 54
4-1 Converting Between Percents, Fractions, and Decimals 55
4-2 Table of Common Percent 56
4-3 Percentage Calculations 57
4-4 Find What Percentage a Number Is of Another 58
4-5 Find a Percentage of a Given Number 59
4-6 Percent Problems 60
4-7 Percent of Increase and Decrease 61
4-8 Markup, Discount, and Tax 62
Answers of Worksheets – Chapter 4 63
Chapter 5: Algebraic Expressions 67
5-1 Expressions and Variables 68
5-2 Simplifying Variable Expressions 69
5-3 The Distributive Property 70
5-4 Translate Phrases into an Algebraic Statement 71
5-5 Evaluating One Variable 72
5-6 Evaluating Two Variables 73

5-7 Combining like Terms 74

5-8 Simplifying Polynomial Expressions 75

Answers of Worksheets – Chapter 5 76

Chapter 6: Equations 79

6-1 One-Step Equations 80

6-2 One-Step Equation Word Problems 81

6-3 Two-Step Equations 82

6-4 Two-Step Equation Word Problems 83

6-5 Multi-Step Equations 84

Answers of Worksheets – Chapter 6 85

Chapter 7: Systems of Equations 87

7-1 Solving Systems of Equations by Graphing 88

7-2 Solving Systems of Equations by Substitution 89

7-3 Solving Systems of Equations by Elimination 90

7-4 Systems of Equations Word Problems 91

Answers of Worksheets – Chapter 7 92

Chapter 8: Inequalities 94

8-1 Graphing Single-Variable Inequalities 95

8-2 One-Step Inequalities 96

8-3 Two-Step Inequalities 97

8-4 Multi-Step Inequalities 98

Answers of Worksheets – Chapter 8 99

Chapter 9: Linear Functions 102

9-1 Finding Slope 103

9-2 Graphing Lines Using Slope-Intercept Form 104

9-3 Graphing Lines Using Standard Form 105

9-4 Writing Linear Equations 106

9-5 Graphing Linear Inequalities 107

Arithmetic and Pre-Algebra Workbook

9-6 Finding Midpoint 108

9-7 Finding Distance of Two Points 109

Answers of Worksheets – Chapter 9 110

Chapter 10: Polynomials 116

10-1 Classifying Polynomials 117

10-2 Writing Polynomials in Standard Form 118

10-3 Simplifying Polynomials 119

10-4 Adding and Subtracting Polynomials 120

10-5 Multiplying Monomials 121

10-6 Multiplying and Dividing Monomials 122

10-7 Multiplying a Polynomial and a Monomial 123

10-8 Multiplying Binomials 124

10-9 Factoring Trinomials 125

10-10 Operations with Polynomials 126

Answers of Worksheets – Chapter 10 127

Chapter 11: Exponents and Radicals 131

11-1 Multiplication Property of Exponents 132

11-2 Division Property of Exponents 133

11-3 Powers of Products and Quotients 134

11-4 Zero and Negative Exponents 135

11-5 Negative Exponents and Negative Bases 136

11-6 Writing Scientific Notation 137

11-7 Square Roots 138

Answers of Worksheets – Chapter 11 139

Chapter 12: Plane Figures 142

12-1 Transformations: Translations, Rotations, and Reflections 143

12-2 The Pythagorean Theorem 144

12-3 Classifying Triangles and Quadrilaterals 145

12-4 Area of Triangles 146

12-5 Perimeter of Polygons 147

12-6 Area and Circumference of Circles 148

12-7 Area of Squares, Rectangles, and Parallelograms 149

12-8 Area of Trapezoids 150

Answers of Worksheets – Chapter 12 152

Chapter 13: Solid Figures 155

13-1 Classifying Solids 156

13-2 Volume of Cubes and Rectangle Prisms 157

13-3 Surface Area of Cubes 158

13-4 Surface Area of a Prism 159

13-5 Surface Area of a Cylinder 160

13-6 Surface Area of Pyramids and Cones 161

13-7 Surface Area of a Sphere 162

13-8 Volume of a Pyramid and Cone 163

13-9 Volume of a Sphere 164

Answers of Worksheets – Chapter 13 165

Chapter 14: Statistics 167

14-1 Mean, Median, Mode, and Range of the Given Data 168

14-2 First Quartile, Second Quartile and Third Quartile of the Given Data 169

14-3 Bar Graph 170

14-4 Box and Whisker Plots 171

14-5 Stem-And-Leaf Plot 172

14-6 The Pie Graph or Circle Graph 173

14-7 Scatter Plots 174

Answers of Worksheets – Chapter 14 175

Chapter 1: Arithmetic

1–1 Simplifying Fractions

1–2 Adding and Subtracting Fractions

1–3 Multiplying and Dividing Fractions

1–4 Adding and Subtracting Mixed Numbers

1–5 Multiplying and Dividing Mixed Numbers

1–6 Adding and Subtracting Decimals

1–7 Multiplying and Dividing Decimals

1–8 Converting Between Fractions, Decimals and Mixed Numbers

1–9 Rounding Numbers

1–10 Factoring Numbers

1–11 Greatest Common Factor

1–12 Least Common Multiple

1-1 Simplifying Fractions

Simplify the fractions.

1) $\dfrac{22}{36}$

2) $\dfrac{8}{10}$

3) $\dfrac{12}{18}$

4) $\dfrac{6}{8}$

5) $\dfrac{13}{39}$

6) $\dfrac{5}{20}$

7) $\dfrac{16}{36}$

8) $\dfrac{18}{36}$

9) $\dfrac{20}{50}$

10) $\dfrac{6}{54}$

11) $\dfrac{45}{81}$

12) $\dfrac{21}{28}$

13) $\dfrac{35}{56}$

14) $\dfrac{52}{64}$

15) $\dfrac{13}{65}$

16) $\dfrac{44}{77}$

17) $\dfrac{21}{42}$

18) $\dfrac{15}{36}$

19) $\dfrac{9}{24}$

20) $\dfrac{20}{80}$

21) $\dfrac{25}{45}$

1-2 Adding and Subtracting Fractions

Add fractions.

1) $\dfrac{2}{3} + \dfrac{1}{2}$

2) $\dfrac{3}{5} + \dfrac{1}{3}$

3) $\dfrac{5}{6} + \dfrac{1}{2}$

4) $\dfrac{7}{4} + \dfrac{5}{9}$

5) $\dfrac{2}{5} + \dfrac{1}{5}$

6) $\dfrac{3}{7} + \dfrac{1}{2}$

7) $\dfrac{3}{4} + \dfrac{2}{5}$

8) $\dfrac{2}{3} + \dfrac{1}{5}$

9) $\dfrac{16}{25} + \dfrac{3}{5}$

Subtract fractions.

10) $\dfrac{4}{5} - \dfrac{2}{5}$

11) $\dfrac{3}{5} - \dfrac{2}{7}$

12) $\dfrac{1}{2} - \dfrac{1}{3}$

13) $\dfrac{8}{9} - \dfrac{3}{5}$

14) $\dfrac{3}{7} - \dfrac{3}{14}$

15) $\dfrac{4}{15} - \dfrac{1}{10}$

16) $\dfrac{3}{4} - \dfrac{13}{18}$

17) $\dfrac{5}{8} - \dfrac{2}{5}$

18) $\dfrac{1}{2} - \dfrac{1}{9}$

1-3 Multiplying and Dividing Fractions

Multiplying fractions. Then simplify.

1) $\dfrac{1}{5} \times \dfrac{2}{3}$

2) $\dfrac{3}{4} \times \dfrac{2}{3}$

3) $\dfrac{2}{5} \times \dfrac{3}{7}$

4) $\dfrac{3}{8} \times \dfrac{1}{3}$

5) $\dfrac{3}{5} \times \dfrac{2}{5}$

6) $\dfrac{7}{9} \times \dfrac{1}{3}$

7) $\dfrac{2}{3} \times \dfrac{3}{8}$

8) $\dfrac{1}{4} \times \dfrac{1}{3}$

9) $\dfrac{5}{7} \times \dfrac{7}{12}$

Dividing fractions.

10) $\dfrac{2}{9} \div \dfrac{1}{4}$

11) $\dfrac{1}{2} \div \dfrac{1}{3}$

12) $\dfrac{6}{11} \div \dfrac{3}{4}$

13) $\dfrac{11}{14} \div \dfrac{1}{10}$

14) $\dfrac{3}{5} \div \dfrac{5}{9}$

15) $\dfrac{1}{2} \div \dfrac{1}{2}$

16) $\dfrac{3}{5} \div \dfrac{1}{5}$

17) $\dfrac{12}{21} \div \dfrac{3}{7}$

18) $\dfrac{5}{14} \div \dfrac{9}{10}$

Arithmetic and Pre-Algebra Workbook

1-4 Adding and Subtracting Mixed Numbers

Add.

1) $4\frac{1}{2} + 5\frac{1}{2}$

2) $2\frac{3}{8} + 3\frac{1}{8}$

3) $5\frac{3}{5} + 5\frac{1}{5}$

4) $1\frac{1}{3} + 2\frac{2}{3}$

5) $5\frac{1}{6} + 5\frac{1}{2}$

6) $3\frac{1}{3} + 1\frac{1}{3}$

7) $1\frac{10}{11} + 1\frac{1}{3}$

8) $2\frac{3}{6} + 1\frac{1}{2}$

9) $5\frac{3}{5} + 5\frac{1}{5}$

10) $7 + \frac{1}{5}$

11) $1\frac{5}{7} + \frac{1}{3}$

12) $2\frac{1}{4} + 1\frac{2}{4}$

Subtract.

13) $4\frac{1}{2} - 3\frac{1}{2}$

14) $3\frac{3}{8} - 3\frac{1}{8}$

15) $6\frac{3}{5} - 5\frac{1}{5}$

16) $2\frac{1}{3} - 1\frac{2}{3}$

17) $6\frac{1}{6} - 5\frac{1}{2}$

18) $3\frac{1}{3} - 1\frac{1}{3}$

19) $2\frac{10}{11} - 1\frac{1}{3}$

20) $2\frac{1}{2} - 1\frac{1}{2}$

21) $6\frac{3}{5} - 2\frac{1}{5}$

22) $7\frac{2}{5} - 1\frac{1}{5}$

23) $2\frac{5}{7} - 1\frac{1}{3}$

24) $2\frac{1}{4} - 1\frac{1}{2}$

Arithmetic and Pre-Algebra Workbook

1-5 Multiplying and Dividing Mixed Numbers

Find each product.

1) $1\frac{2}{3} \times 1\frac{1}{4}$

2) $1\frac{3}{5} \times 1\frac{2}{3}$

3) $1\frac{2}{3} \times 3\frac{2}{7}$

4) $4\frac{1}{8} \times 1\frac{2}{5}$

5) $2\frac{2}{5} \times 3\frac{1}{5}$

6) $1\frac{1}{3} \times 1\frac{2}{3}$

7) $1\frac{5}{8} \times 2\frac{1}{2}$

8) $3\frac{2}{5} \times 2\frac{1}{5}$

9) $2\frac{2}{3} \times 4\frac{1}{4}$

10) $2\frac{3}{5} \times 1\frac{2}{4}$

11) $1\frac{1}{3} \times 1\frac{1}{4}$

12) $3\frac{2}{5} \times 1\frac{1}{5}$

Find each quotient.

13) $2\frac{1}{5} \div 2\frac{1}{2}$

14) $2\frac{3}{5} \div 1\frac{1}{3}$

15) $3\frac{1}{6} \div 4\frac{2}{3}$

16) $1\frac{2}{3} \div 3\frac{1}{3}$

17) $4\frac{1}{8} \div 2\frac{2}{4}$

18) $3\frac{1}{2} \div 2\frac{3}{5}$

19) $3\frac{5}{9} \div 1\frac{2}{5}$

20) $2\frac{2}{7} \div 1\frac{1}{2}$

21) $3\frac{1}{5} \div 1\frac{1}{2}$

22) $4\frac{3}{5} \div 2\frac{1}{3}$

23) $6\frac{1}{6} \div 1\frac{2}{3}$

24) $2\frac{2}{3} \div 1\frac{1}{3}$

1-6 Adding and Subtracting Decimals

Add and subtract decimals.

1) 15.14
   $-\ 12.18$
   _____

2) 65.72
   $+\ 43.67$
   _____

3) 82.56
   $+\ 12.28$
   _____

4) 34.18
   $-\ 23.45$
   _____

5) 90.37
   $+\ 56.97$
   _____

6) 45.78
   $-\ 23.39$
   _____

Solve.

7) ____ + 1.3 = 4.8

8) 4.2 + ____ = 11.6

9) 9.9 + ____ = 16

10) 6.9 + ____ = 16.4

11) ____ + 5.1 = 8.6

12) ____ + 7.9 = 15.2

1-7 Multiplying and Dividing Decimals

Find each product.

1) 4.5 × 1.6

2) 7.7 × 9.9

3) 2.6 × 1.5

4) 8.9 × 9.7

5) 15.1 × 12.6

6) 6.9 × 3.3

7) 5.7 × 7.8

8) 98.20 × 100

9) 23.99 × 1000

Find each quotient.

1) 9.2 ÷ 3.6

2) 27.6 ÷ 3.8

3) 12.6 ÷ 4.7

4) 6.5 ÷ 8.1

5) 1.4 ÷ 10

6) 3.6 ÷ 100

7) 4.24 ÷ 10

8) 14.6 ÷ 100

1-8 Converting Between Fractions, Decimals and Mixed Numbers

Convert fractions to decimals.

1) $\dfrac{9}{10}$

2) $\dfrac{56}{100}$

3) $\dfrac{3}{4}$

4) $\dfrac{2}{5}$

5) $\dfrac{1}{3}$

6) $\dfrac{4}{5}$

7) $\dfrac{5}{6}$

8) $\dfrac{5}{8}$

9) $\dfrac{13}{21}$

Convert decimal into fraction.

10) 0.3

11) 4.5

12) 2.5

13) 2.3

14) 0.8

15) 0.25

16) 0.14

17) 0.2

18) 0.08

19) 0.45

20) 2.6

21) 5.2

1-9 Rounding Numbers

Round each number to the underlined place value.

1) 9̲72

2) 2,9̲95

3) 36̲4

4) 8̲1

5) 5̲5

6) 33̲4

7) 1,2̲03

8) 9.5̲7

9) 7.4̲84

10) 9.1̲4

11) 3̲9

12) 9̲,123

13) 3,45̲2

14) 5̲69

15) 1,2̲30

16) 9̲8

17) 9̲3

18) 3̲7

19) 49̲3

20) 2,9̲23

21) 9̲,845

22) 55̲5

23) 9̲39

24) 6̲9

1-10 Factoring Numbers

List all positive factors of each number.

1) 68

2) 56

3) 24

4) 40

5) 86

6) 78

7) 50

8) 98

9) 45

10) 26

11) 54

12) 28

13) 55

14) 85

15) 50

List the prime factorization for each number.

16) 50

17) 25

18) 69

19) 21

20) 45

21) 68

22) 26

23) 86

24) 93

1-11 Greatest Common Factor

Find the GCF for each number pair.

1) 20, 30

2) 4, 14

3) 5, 45

4) 68, 12

5) 5, 12

6) 15, 27

7) 3, 24

8) 34, 6

9) 4, 10

10) 5, 3

11) 6, 16

12) 30, 3

13) 24, 28

14) 70, 10

15) 45, 8

16) 90, 35

17) 78, 34

18) 55, 75

19) 60, 72

20) 100, 78

21) 30, 40

1-12 Least Common Multiple

Find the LCM for each number pair.

1) 4, 14

2) 5, 15

3) 16, 10

4) 4, 34

5) 8, 3

6) 12, 24

7) 9, 18

8) 5, 6

9) 8, 19

10) 9, 21

11) 19, 29

12) 7, 6

13) 25, 6

14) 4, 8

15) 30, 10, 50

16) 18, 36, 27

17) 12, 8, 18

18) 8, 18, 4

19) 26, 20, 30

20) 10, 4, 24

21) 15, 30, 45

Answers of Worksheets – Chapter 1

1–1 Simplifying Fractions

1) $\dfrac{11}{18}$
2) $\dfrac{4}{5}$
3) $\dfrac{2}{3}$
4) $\dfrac{3}{4}$
5) $\dfrac{1}{3}$
6) $\dfrac{1}{4}$
7) $\dfrac{4}{9}$
8) $\dfrac{1}{2}$
9) $\dfrac{2}{5}$
10) $\dfrac{1}{9}$
11) $\dfrac{5}{9}$
12) $\dfrac{3}{4}$
13) $\dfrac{5}{8}$
14) $\dfrac{13}{16}$
15) $\dfrac{1}{5}$
16) $\dfrac{4}{7}$
17) $\dfrac{1}{2}$
18) $\dfrac{5}{12}$
19) $\dfrac{3}{8}$
20) $\dfrac{1}{4}$
21) $\dfrac{5}{9}$

1–2 Adding and Subtracting Fractions

1) $\dfrac{7}{6}$
2) $\dfrac{14}{15}$
3) $\dfrac{4}{3}$
4) $\dfrac{83}{36}$
5) $\dfrac{3}{5}$
6) $\dfrac{13}{14}$
7) $\dfrac{23}{20}$
8) $\dfrac{13}{15}$
9) $\dfrac{31}{25}$
10) $\dfrac{2}{5}$
11) $\dfrac{11}{35}$
12) $\dfrac{1}{6}$
13) $\dfrac{13}{45}$
14) $\dfrac{3}{14}$
15) $\dfrac{1}{6}$
16) $\dfrac{1}{36}$
17) $\dfrac{9}{40}$
18) $\dfrac{7}{18}$

Arithmetic and Pre-Algebra Workbook

1–3 Multiplying and Dividing Fractions

1) $\frac{2}{15}$
2) $\frac{1}{2}$
3) $\frac{6}{35}$
4) $\frac{1}{8}$
5) $\frac{6}{25}$
6) $\frac{7}{27}$
7) $\frac{1}{4}$
8) $\frac{1}{12}$
9) $\frac{5}{12}$
10) $\frac{8}{9}$
11) $\frac{3}{2}$
12) $\frac{8}{11}$
13) $\frac{55}{7}$
14) $\frac{27}{25}$
15) 1
16) 3
17) $\frac{4}{3}$
18) $\frac{25}{63}$

1–4 Adding and Subtracting Mixed Numbers

1) 10
2) $5\frac{1}{2}$
3) $10\frac{4}{5}$
4) 4
5) $10\frac{2}{3}$
6) $4\frac{2}{3}$
7) $3\frac{8}{33}$
8) 4
9) $10\frac{4}{5}$
10) $7\frac{1}{5}$
11) $2\frac{1}{21}$
12) $3\frac{3}{4}$
13) 1
14) $\frac{1}{4}$
15) $1\frac{2}{5}$
16) $\frac{2}{3}$
17) $\frac{2}{3}$
18) 2
19) $1\frac{19}{33}$
20) 1
21) $4\frac{2}{5}$
22) $6\frac{1}{5}$
23) $1\frac{8}{21}$
24) $\frac{3}{4}$

Arithmetic and Pre-Algebra Workbook

1–5 Multiplying and Dividing Mixed Numbers

1) $2\frac{1}{12}$
2) $2\frac{2}{3}$
3) $5\frac{10}{21}$
4) $5\frac{31}{40}$
5) $7\frac{17}{25}$
6) $2\frac{2}{9}$
7) $4\frac{1}{16}$
8) $7\frac{12}{25}$
9) $11\frac{1}{3}$
10) $3\frac{9}{10}$
11) $1\frac{2}{3}$
12) $4\frac{2}{25}$
13) $\frac{22}{25}$
14) $1\frac{19}{20}$
15) $\frac{19}{28}$
16) $\frac{1}{2}$
17) $1\frac{13}{20}$
18) $1\frac{9}{26}$
19) $2\frac{34}{63}$
20) $1\frac{11}{20}$
21) $2\frac{2}{15}$
22) $1\frac{34}{35}$
23) $3\frac{7}{10}$
24) 2

1–6 Adding and Subtracting Decimals

1) 2.96
2) 109.39
3) 94.84
4) 10.73
5) 147.34
6) 22.39
7) 3.5
8) 7.4
9) 6.1
10) 9.5
11) 3.5
12) 7.3

1–7 Multiplying and Dividing Decimals

1) 7.2
2) 76.23
3) 3.9
4) 86.33
5) 190.26
6) 22.77
7) 44.46
8) 9820
9) 23990
10) 2.5555…
11) 7.2631…
12) 2.6808…
13) 0.8024…
14) 0.14
15) 0.036
16) 0.424
17) 0.146

1–8 Converting Between Fractions, Decimals and Mixed Numbers

1) 0.9
2) 0.56
3) 0.75
4) 0.4
5) 0.333...
6) 0.8
7) 0.8333...
8) 0.625
9) 0.6190...
10) $\frac{3}{10}$
11) $4\frac{1}{2}$
12) $2\frac{1}{2}$
13) $2\frac{3}{10}$
14) $\frac{4}{5}$
15) $\frac{1}{4}$
16) $\frac{7}{50}$
17) $\frac{1}{5}$
18) $\frac{2}{25}$
19) $\frac{9}{20}$
20) $2\frac{3}{5}$
21) $5\frac{1}{5}$

1–9 Rounding Numbers

1) 1,000
2) 3,000
3) 360
4) 80
5) 60
6) 330
7) 1,200
8) 9.6
9) 7.5
10) 9.1
11) 40
12) 9,000
13) 3,450
14) 600
15) 1,200
16) 100
17) 90
18) 40
19) 490
20) 2,900
21) 10,000
22) 560
23) 900
24) 70

1–10 Factoring Numbers

1) 1, 2, 4, 17, 34, 68
2) 1, 2, 4, 7, 8, 14, 28, 56
3) 1, 2, 3, 4, 6, 8, 12, 24
4) 1, 2, 4, 5, 8, 10, 20, 40
5) 1, 2, 43, 86
6) 1, 2, 3, 6, 13, 26, 39, 78
7) 1, 2, 5, 10, 25, 50
8) 1, 2, 7, 14, 49, 98
9) 1, 3, 5, 9, 15, 45
10) 1, 2, 13, 26
11) 1, 2, 3, 6, 9, 18, 27, 54
12) 1, 2, 4, 7, 14, 28
13) 1, 5, 11, 55
14) 1, 5, 17, 85

15) 1, 2, 5, 10, 25, 50
16) 2 × 5 × 5
17) 5 × 5
18) 3 × 23
19) 3 × 7

20) 3 × 3 × 5
21) 2 × 2 × 17
22) 2 × 13
23) 2 × 43
24) 3 × 31

1–11 Greatest Common Factor

1) 10
2) 2
3) 5
4) 4
5) 1
6) 3
7) 3

8) 2
9) 2
10) 1
11) 2
12) 3
13) 4
14) 10

15) 1
16) 5
17) 2
18) 5
19) 12
20) 2
21) 10

1–12 Least Common Multiple

1) 28
2) 15
3) 80
4) 68
5) 24
6) 24
7) 18

8) 30
9) 152
10) 63
11) 551
12) 42
13) 150
14) 8

15) 150
16) 108
17) 72
18) 72
19) 780
20) 120
21) 90

Chapter 2: Real Numbers and Integers

2–1 Adding and Subtracting Integers

2–2 Multiplying and Dividing Integers

2–3 Ordering Integers and Numbers

2–4 Arrange and Order, Comparing Integers

2–5 Order of Operations

2–6 Mixed Integer Computations

2–7 Absolute Value

2–8 Integers and Absolute Value

2–9 Classifying Real Numbers Venn Diagram

2-1 Adding and Subtracting Integers

Find the sum.

1) $(-12) + (-4)$

2) $5 + (-24)$

3) $(-14) + 23$

4) $(-8) + (39)$

5) $43 + (-12)$

6) $(-23) + (-4) + 3$

7) $4 + (-12) + (-10) + (-25)$

8) $19 + (-15) + 25 + 11$

9) $(-9) + (-12) + (32 - 14)$

10) $4 + (-30) + (45 - 34)$

Find the difference.

11) $(-14) - (-9) - (18)$

12) $(-9) - (-25)$

13) $(-12) - (8)$

14) $(28) - (4)$

15) $(34) - (2)$

16) $(55) - (-5) + (-4)$

17) $(9) - (2) - (-5)$

18) $(2) - (4) - (-15)$

19) $(23) - (4) - (-34)$

20) $(-45) - (-87)$

2-2 Multiplying and Dividing Integers

Find each product.

1) $(-8) \times (-2)$

2) 3×6

3) $(-4) \times 5 \times (-6)$

4) $2 \times (-6) \times (-6)$

5) $11 \times (-12)$

6) $10 \times (-5)$

7) 8×8

8) $(-8) \times (-9)$

9) $6 \times (-5) \times 3$

10) $6 \times (-1) \times 2$

Find each quotient.

11) $18 \div 3$

12) $(-24) \div 4$

13) $(-63) \div (-9)$

14) $54 \div 9$

15) $20 \div (-2)$

16) $(-66) \div (-11)$

17) $64 \div 8$

18) $(-121) \div 11$

19) $72 \div 9$

20) $16 \div 4$

2-3 Ordering Integers and Numbers

Order each set of integers from least to greatest.

1) $-15, -19, 20, -4, 1$ ___, ___, ___, ___, ___, ___

2) $6, -5, 4, -3, 2$ ___, ___, ___, ___, ___, ___

3) $15, -42, 19, 0, -22$ ___, ___, ___, ___, ___, ___

4) $26, -91, 0, -13, 67, -55$ ___, ___, ___, ___, ___, ___

5) $-17, -71, 90, -25, -54, -39$ ___, ___, ___, ___, ___, ___

6) $98, 5, 46, 19, 77, 24$ ___, ___, ___, ___, ___, ___

Order each set of integers from greatest to least.

7) $-2, 5, -3, 6, -4$ ___, ___, ___, ___, ___, ___

8) $-37, 7, -17, 27, 47$ ___, ___, ___, ___, ___, ___

9) $32, -27, 19, -17, 15$ ___, ___, ___, ___, ___, ___

10) $68, 81, 21, -18, 94, 72$ ___, ___, ___, ___, ___, ___

2-4 Arrange, Order, and Comparing Integers

Arrange these integers in descending order.

1) 21, 71, – 18, – 10, 82   ___, ___, ___, ___, ___

2) 15, 11, 20, 12, – 9, – 5   ___, ___, ___, ___, ___, ___

3) – 5, 20, 15, 9, –11   ___, ___, ___, ___, ___

4) 19, 18, – 9, – 6, – 11   ___, ___, ___, ___, ___

5) 56, – 34, – 12, – 5, 32   ___, ___, ___, ___, ___

Compare. Use >, =, <

6) – 8 ____ 12   11) – 56 ____ – 58

7) – 10 ____ –16   12) 78 ____ 87

8) 43 ____ 34   13) – 92 ____ – 102

9) 15 ____ –16   14) – 12 ____ – 12

10) – 354 ____ –345   15) – 721 ____ – 821

2-5 Order of Operations

Evaluate each expression.

1) $(2 \times 2) + 5$

2) $24 - (3 \times 3)$

3) $(6 \times 4) + 8$

4) $25 - (4 \times 2)$

5) $(6 \times 5) + 3$

6) $64 - (2 \times 4)$

7) $25 + (1 \times 8)$

8) $(6 \times 7) + 7$

9) $48 \div (4 + 4)$

10) $(7 + 11) \div (-2)$

11) $9 + (2 \times 5) + 10$

12) $(5 + 8) \times \dfrac{3}{5} + 2$

13) $2 \times 7 - \left(\dfrac{10}{9 - 4}\right)$

14) $(12 + 2 - 5) \times 7 - 1$

15) $\left(\dfrac{7}{5 - 1}\right) \times (2 + 6) \times 2$

16) $20 \div (4 - (10 - 8))$

17) $\dfrac{50}{4\,(5 - 4) - 3}$

18) $2 + (8 \times 2)$

2-6 Mixed Integer Computations

Compute.

1) $(-70) \div (-5)$

2) $(-14) \times 3$

3) $(-4) \times (-15)$

4) $(-65) \div 5$

5) $18 \times (-7)$

6) $(-12) \times (-2)$

7) $\dfrac{(-60)}{(-20)}$

8) $24 \div (-8)$

9) $22 \div (-11)$

10) $\dfrac{(-27)}{3}$

11) $4 \times (-4)$

12) $\dfrac{(-48)}{12}$

13) $(-14) \times (-2)$

14) $(-7) \times (7)$

15) $\dfrac{-30}{-6}$

16) $(-54) \div 6$

17) $(-60) \div (-5)$

18) $(-7) \times (-12)$

19) $(-14) \times 5$

20) $88 \div (-8)$

2-7 Absolute Value

Evaluate.

1) $|-4| + |-12| - 7$

2) $|-5| + |-13|$

3) $-18 + |-5 + 3| - 8$

4) $|27| \div |9|$

5) $|-9| \div |-1|$

6) $|200| \div |-100|$

7) $|55| \div |11|$

8) $|36| \div |-6|$

9) $|25| \times |-5|$

10) $|-3| \times |-8|$

11) $|12| \times |-5|$

12) $|11| \times |-6|$

13) $|-8| \times |4|$

14) $|-9| \times |-7|$

15) $|43 - 67 + 9| + |-11| - 1$

16) $|-45 + 78| + |23| - |45|$

17) $75 + |-11 - 30| - |2|$

18) $|-3 + 15| + |9 + 4| - 1$

2-8 Integers and Absolute Value

Write absolute value of each number.

1) -4

2) -7

3) -8

4) 4

5) 5

6) -10

7) 1

8) 6

9) 8

10) -2

11) -1

12) 10

13) 3

14) 7

15) -5

16) -3

17) -9

18) 2

19) 4

20) -6

21) 9

Evaluate.

22) $|-43| - |12| + 10$

23) $76 + |-15 - 45| - |3|$

24) $30 + |-62| - 46$

25) $|32| - |-78| + 90$

26) $|-35 + 4| + 6 - 4$

27) $|-4| + |-11|$

28) $|-6 + 3 - 4| + |7 + 7|$

29) $|-9| + |-19| - 5$

2-9 Classifying Real Numbers Venn Diagram

Identify all of the subsets of real number system to which each number belongs:

Example:

0.1259 : Rational number

$\sqrt{2}$: Irrational number

3 : Natural number, whole number, Integer, rational number

1) 0

2) -5

3) -8.5

4) $\sqrt{4}$

5) -10

6) 18

7) 6

8) π

9) $1\frac{2}{7}$

10) -1

11) $\sqrt{5}$

Answers of Worksheets – Chapter 2

2–1 Adding and Subtracting Integers

1) −16
2) −19
3) 9
4) 31
5) 31
6) −24
7) −43
8) 40
9) −3
10) −15
11) −23
12) 16
13) −20
14) 32
15) 32
16) 56
17) 12
18) 13
19) 53
20) 42

2–2 Multiplying and Dividing Integers

1) 16
2) 18
3) 120
4) 72
5) −132
6) −50
7) 64
8) 72
9) −90
10) −12
11) 6
12) −6
13) 7
14) 6
15) −10
16) 6
17) 8
18) −11
19) 8
20) 4

2–3 Ordering Integers and Numbers

1) −19, −15, −4, 1, 20
2) −5, −3, 2, 4, 6
3) −42, −22, 0, 15, 19
4) −91, −55, −13, 0, 26, 67
5) −71, −54, −39, −25, −17, 90
6) 5, 19, 24, 46, 77, 98
7) 6, 5, −2, −3, −4
8) 47, 27, 7, −17, −37
9) 32, 19, 15, −17, −27
10) 94, 81, 72, 68, 21, −18

Arithmetic and Pre-Algebra Workbook

2–4 Arrange and Order, Comparing Integers

1) 82, 71, 21, − 10, − 18
2) 20, 15, 12, 11, − 5, − 9
3) 20, 15, 9, − 5, −11
4) 19, 18, − 6, − 9, − 11
5) 56, 32, − 5, − 12, − 34
6) <
7) >
8) >
9) >
10) <
11) >
12) <
13) >
14) =
15) >

2–5 Order of Operations

1) 9
2) 15
3) 32
4) 17
5) 33
6) 56
7) 33
8) 49
9) 6
10) − 9
11) 29
12) 9.8
13) 12
14) 62
15) 28
16) 10
17) 50
18) 18

2–6 Mixed Integer Computations

1) 14
2) − 42
3) 60
4) − 13
5) − 126
6) 24
7) 3
8) − 3
9) − 2
10) − 9
11) − 16
12) − 4
13) 28
14) − 49
15) 5
16) − 9
17) 12
18) 84
19) − 70
20) − 11

2–7 Absolute Value

1) 9
2) 18
3) − 24
4) 3
5) 9
6) 2
7) 5
8) 6
9) 125
10) 24
11) 60
12) 66
13) 32
14) 63
15) 25

www.EffortlessMath.com

Arithmetic and Pre-Algebra Workbook

16) 11   17) 114   18) 24

2–8 Integers and Absolute Value

1) 4
2) 7
3) 8
4) 4
5) 5
6) 10
7) 1
8) 6
9) 8
10) 2
11) 1
12) 10
13) 3
14) 7
15) 5
16) 3
17) 9
18) 2
19) 4
20) 6
21) 9
22) 41
23) 133
24) 46
25) 44
26) 33
27) 15
28) 21
29) 23

2–9 Classifying Real Numbers Venn Diagram

1) 0: whole number, integer, rational number
2) -5: integer, rational number
3) -8.5: rational number
4) $\sqrt{4}$: natural number, whole number, integer, rational number
5) -10: integer, rational number
6) 18 : natural number, whole number, integer, rational number
7) 6: natural number, whole number, integer, rational number
8) π: irrational number
9) $1\frac{2}{7}$: rational number
10) -1: integer, rational number
11) $\sqrt{5}$: irrational number

Chapter 3: Proportions and Ratios

3–1 Writing Ratios

3–2 Simplifying Ratios

3–3 Proportional Ratios

3–4 Create a Proportion

3–5 Similar Figures

3–6 Similar Figure Word Problems

3–7 Simple and Compound Interest

3–8 Complete the Ratio Table

3–9 Ratio and Rates Word Problems

3-1 Writing Ratios

Express each ratio as a rate and unite rate.

1) 12 miles on 4 gallons of gas.

2) 24 dollars for 6 books.

3) 200 miles on 14 gallons of gas

4) 24 inches of snow in 8 hours

Express each ratio as a fraction in the simplest form.

5) 3 feet out of 30 feet

6) 18 cakes out of 42 cakes

7) 16 dimes t0 24 dimes

8) 12 dimes out of 48 coins

9) 14 cups to 84 cups

10) 45 gallons to 65 gallons

11) 10 miles out of 40 miles

12) 22 blue cars out of 55 cars

13) 32 pennies to 300 pennies

14) 24 beetles out of 86 insects

3-2 Simplifying Ratios

Reduce each ratio.

1) 21 : 49

2) 20 : 40

3) 10 : 50

4) 14 : 18

5) 45 : 27

6) 49 : 21

7) 100 : 10

8) 12 : 8

9) 35 : 45

10) 8 : 20

11) 25 : 35

12) 21 : 27

13) 52 : 82

14) 12 : 36

15) 24 : 3

16) 15 : 30

17) 3 : 36

18) 8 : 16

19) 6 : 100

20) 2 : 20

21) 10 : 60

22) 14 : 63

23) 68 : 80

24) 8 : 80

Arithmetic and Pre-Algebra Workbook

3-3 Proportional Ratios

Solve each proportion.

1) $\dfrac{3}{6} = \dfrac{8}{d}$

2) $\dfrac{k}{5} = \dfrac{120}{15}$

3) $\dfrac{30}{5} = \dfrac{12}{x}$

4) $\dfrac{x}{2} = \dfrac{80}{10}$

5) $\dfrac{d}{3} = \dfrac{2}{6}$

6) $\dfrac{27}{7} = \dfrac{30}{x}$

7) $\dfrac{8}{5} = \dfrac{k}{15}$

8) $\dfrac{60}{20} = \dfrac{3}{d}$

9) $\dfrac{x}{3} = \dfrac{12}{18}$

10) $\dfrac{25}{5} = \dfrac{x}{8}$

11) $\dfrac{12}{x} = \dfrac{4}{2}$

12) $\dfrac{x}{4} = \dfrac{18}{2}$

13) $\dfrac{80}{10} = \dfrac{k}{10}$

14) $\dfrac{12}{6} = \dfrac{6}{d}$

15) $\dfrac{x}{4} = \dfrac{30}{5}$

16) $\dfrac{9}{5} = \dfrac{k}{5}$

17) $\dfrac{45}{15} = \dfrac{15}{d}$

18) $\dfrac{60}{x} = \dfrac{10}{3}$

19) $\dfrac{d}{3} = \dfrac{14}{6}$

20) $\dfrac{k}{4} = \dfrac{4}{2}$

21) $\dfrac{4}{2} = \dfrac{x}{7}$

3-4 Create a Proportion

Create proportion from the given set of numbers.

1) 1, 6, 2, 3

2) 12, 144, 1, 12

3) 16, 4, 8, 2

4) 9, 5, 27, 15

5) 7, 10, 60, 42

6) 8, 7, 24, 21

7) 10, 5, 8, 4

8) 3, 12, 8, 2

9) 2, 2, 1, 4

10) 3, 6, 7, 14

11) 2, 6, 5, 15

12) 7, 2, 14, 4

www.EffortlessMath.com

3-5 Similar Figures

Each pair of figures is similar. Find the value of x.

1)

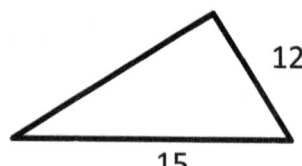

2)

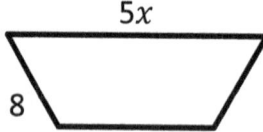

3)

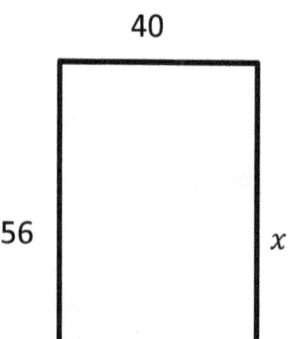

3-6 Similar Figure Word Problems

Answer each question and round your answer to the nearest whole number.

1) If a 42.9 ft tall flagpole casts a 253.1 ft long shadow, then how long is the shadow that a 6.2 ft tall woman casts?

2) A model igloo has a scale of 1 in : 2 ft. If the real igloo is 10 ft wide then how wide is the model igloo?

3) If a 18 ft tall tree casts a 9 ft long shadow, then how tall is an adult giraffe that casts a 7 ft shadow?

4) Find the distance between San Joe and Mount Pleasant if they are 2 cm apart on a map with a scale of 1 cm : 9 km.

5) A telephone booth that is 8 ft tall casts a shadow that is 4 ft long. Find the height of a lawn ornament that casts a 2 ft shadow.

www.EffortlessMath.com

3-7 Simple Interest

Use simple interest to find the ending balance.

1) $1,300 at 5% for 6 years.

2) $5,400 at 7.5% for 6 months.

3) $25,600 at 9.2% for 5 years.

4) $24,000 at 8.5% for 9 years.

5) $240 interest is earned on a principal of $1500 at a simple interest rate of 4% interest per year. For how many years was the principal invested?

6) A new car, valued at $28,000, depreciates at 9% per year from original price. Find the value of the car 3 years after purchase.

7) Sara puts $2,000 into an investment yielding 5% annual simple interest; she left the money in for five years. How much interest does Sara get at the end of those five years?

Arithmetic and Pre-Algebra Workbook

3-8 Complete the Ratio Table

Complete the ratio tables.

1)

6		18	
7	14		28

2)

3		9	12
7	14		

3)

10		30	40
5	10		20

4)

1	4	5	9
2			

5)

8		40	48
5	10	25	30

6)

2	4	8	10
3			

7)

11		33	44
15	30		

8)

2	6	8	10
5			

9)

8		40	48
5	10	25	30

10)

3	9	21	27
			36

www.EffortlessMath.com

3-9 Ratio and Rates Word Problems

Solve.

1) In a party, 10 soft drinks are required for every 12 guests. If there are 252 guests, how many soft drink is required?

2) In Jack's class, 18 of the students are tall and 10 are short. In Michael's class 54 students are tall and 30 students are short. Which class has a higher ratio of tall to short students?

3) Are these ratios equivalent?
   12 cards to 72 animals, 11 marbles to 66 marbles

4) The price of 3 apples at the Quick Market is $1.44. The price of 5 of the same apples at Walmart is $2.50. Which place is the better buy?

5) The bakers at a Bakery can make 160 bagels in 4 hours. How many bagels can they bake in 16 hours? What is that rate per hour?

6) You can buy 5 cans of green beans at a supermarket for $3.40. How much does it cost to buy 35 cans of green beans?

Arithmetic and Pre-Algebra Workbook

Answers of Worksheets – Chapter 3

3–1 Writing Ratios

1) $\frac{120\ miles}{4\ gallons}$, 30 miles per gallon
2) $\frac{36\ dollars}{6\ books}$, 6.00 doller book
3) $\frac{200\ miles}{14\ gallons}$, 14.29 miles per gallon
4) $\frac{24"\ of\ snow}{8\ hours}$, 3 inches of snow per hour

5) $\frac{1}{10}$
6) $\frac{3}{7}$
7) $\frac{2}{3}$
8) $\frac{1}{4}$
9) $\frac{1}{6}$
10) $\frac{9}{13}$
11) $\frac{1}{4}$
12) $\frac{2}{5}$
13) $\frac{8}{75}$
14) $\frac{12}{43}$

3–2 Simplifying Ratios

1) 3 : 7
2) 1 : 2
3) 1 : 5
4) 7 : 9
5) 5 : 3
6) 7 : 3
7) 10 : 1
8) 3 : 2
9) 7 : 9
10) 2 : 5
11) 5 : 7
12) 7 : 9
13) 26 : 41
14) 1 : 3
15) 8 : 1
16) 1 : 2
17) 1 : 12
18) 1 : 2
19) 3 : 50
20) 1 : 10
21) 1 : 6
22) 2 : 9
23) 17 : 20
24) 1 : 10

3–3 Proportional Ratios

1) 16
2) 4
3) 2
4) 0.25
5) 1
6) 7.78
7) 24
8) 1
9) 2
10) 40
11) 6
12) 36
13) 80
14) 3
15) 24
16) 9
17) 5
18) 18

www.EffortlessMath.com

Arithmetic and Pre-Algebra Workbook

19) 7   20) 8   21) 14

3–4 Create a Proportion

1) 1 : 3 = 2 : 6   5) 7 : 42, 10 : 60   9) 4 : 2 = 2 : 1
2) 12 : 144 = 1 : 12   6) 7 : 21 = 8 : 24   10) 7 : 3 = 14 : 6
3) 2 : 4 = 8 : 16   7) 8 : 10 = 4 : 5   11) 5 : 2 = 15 : 6
4) 5 : 15 = 9 : 27   8) 2 : 3 = 8 : 12   12) 7 : 2 = 14 : 4

3–5 Similar Figures

1) 5   2) 3   3) 56

3–6 Similar Figure Word Problems

1) 36.6 ft   3) 14 ft   5) 4 ft
2) 5 in   4) 18 km

3–7 Simple and Compound Interest

1) $1,690.00   4) $42,360.00   7) $500.00
2) $5,602.50   5) 4 years
3) $37,376.00   6) $20,440

3–8 Complete the Ratio Table

1)

6	12	18	24
7	14	21	28

2)

3	6	9	12
7	14	21	28

3)

10	20	30	40
5	10	15	20

6)

2	4	8	10
3	6	12	15

7)

11	22	33	44
15	30	45	60

8)

2	6	8	10
5	15	20	25

4)

1	4	5	9
2	8	10	18

5)

8	16	40	48
5	10	25	30

9)

8	16	40	48
5	10	25	30

10)

3	9	21	27
4	12	28	36

3–9 Ratio and Rates Word Problems

1) 210

2) The ratio for both class is equal to 9 to 5.

3) Yes! Both ratios are 1 to 6.

4) The price at the Quick Market is a better buy.

5) 640, the rate is 40 per hour.

6) $23.80

Chapter 4: Percent

4–1 Converting Between Percent, Fractions, and Decimals

4–3 Percentage Calculations

4–4 Find What Percentage a Number Is of Another

4–5 Find a Percentage of a Given Number

4–6 Percent Problems

4–7 Percent of Increase and Decrease

4–8 Markup, Discount, and Tax

4-1 Converting Between Percents, Fractions, and Decimals

Converting fractions to decimals.

1) $\dfrac{50}{100}$ 	 4) $\dfrac{80}{100}$ 	 7) $\dfrac{90}{100}$

2) $\dfrac{38}{100}$ 	 5) $\dfrac{7}{100}$ 	 8) $\dfrac{20}{100}$

3) $\dfrac{15}{100}$ 	 6) $\dfrac{35}{100}$ 	 9) $\dfrac{7}{100}$

Write each decimal as a percent.

10) 0.5 	 13) 0.524 	 16) 3.63

11) 0.9 	 14) 0.1 	 17) 0.008

12) 0.002 	 15) 0.03 	 18) 4.78

4-2 Table of Common Percent

Complete the table of common percent.

Fraction	Decimal	Percent
$\frac{1}{25}$	0.04	4%
$\frac{1}{2}$		50%
$\frac{1}{4}$	0.25	25%
$\frac{1}{5}$		20%
$\frac{6}{10}$	0.6	
$\frac{5}{8}$		62.5 %
$\frac{2}{5}$		40%
$\frac{7}{100}$	0.07	7%
$\frac{7}{16}$		43.75%
$\frac{5}{8}$		$62\frac{1}{2}$%
$\frac{7}{10}$	0.7	
$\frac{30}{100}$		30%
$\frac{4}{8}$	0.8	
$\frac{3}{4}$		75%

Arithmetic and Pre-Algebra Workbook

4-3 Percentage Calculations

Calculate the percentages.

1) 50% of 25

2) 80% of 15

3) 30% of 34

4) 70% of 45

5) 10% of 0

6) 80% of 22

7) 65% of 8

8) 78% of 54

9) 50% of 80

10) 20% of 10

11) 40% of 40

12) 90% of 0

13) 20% of 70

14) 55% of 60

15) 80% of 10

16) 20% of 880

17) 70% of 100

18) 80% of 90

Solve.

19) 50 is what percentage of 75?
20) What percentage of 100 is 70
21) Find what percentage of 60 is 35.
22) 40 is what percentage of 80?

4-4 Find What Percentage a Number Is of Another

Find the percentage of the numbers.

1) 45 is what percent of 90?

2) 15 is what percent of 75?

3) 20 is what percent of 400?

4) 18 is what percent of 90?

5) 3 is what percent of 15?

6) 8 is what percent of 80?

7) 11 is what percent of 55?

8) 9 is what percent of 90?

9) 2.5 is what percent of 10?

10) 5 is what percent of 25?

11) 60 is what percent of 20?

12) 12 is what percent of 48?

13) 14 is what percent of 28?

14) 8.2 is what percent of 32.8?

15) 1200 is what percent of 4,800?

16) 4,000 is what percent of 20,000?

17) 45 is what percent of 900?

18) 10 is what percent of 200?

19) 15 is what percent of 60?

20) 1.2 is what percent of 24?

4-5 Find a Percentage of a Given Number

Find a Percentage of a Given Number.

1) 90% of 50

2) 40% of 50

3) 10% of 0

4) 80% of 80

5) 60% of 40

6) 50% of 60

7) 30% of 20

8) 35% of 10

9) 10% of 80

10) 10% of 60

11) 100% 0f 50

12) 90% of 34

13) 80% of 42

14) 90% of 12

15) 20% of 56

16) 40% of 40

17) 40% of 6

18) 70% of 38

19) 30% of 3

20) 40% of 50

21) 100% of 8

4-6 Percent Problems

Solve each problem.

1) 51 is 340% of what?

2) 93% of what number is 97?

3) 27% of 142 is what number?

4) What percent of 125 is 29.3?

5) 60 is what percent of 126?

6) 67 is 67% of what?

7) 67 is 13% of what?

8) 41% of 78 is what?

9) 1 is what percent of 52.6?

10) What is 59% of 14 m?

11) What is 90% of 130 inches?

12) 16 inches is 35% of what?

13) 90% of 54.4 hours is what?

14) What percent of 33.5 is 21?

15) Liam scored 22 out of 30 marks in Algebra, 35 out of 40 marks in science and 89 out of 100 marks in mathematics. In which subject his percentage of marks in best?

16) Ella require 50% to pass her test. If she gets 280 marks and falls short by 20 marks, what were the maximum marks she could have got?

4-7 Percent of Increase and Decrease

Find each percent change to the nearest percent. Increase or Decrease.

1) From 32 grams to 82 grams.

2) From 150 m to 45 m

3) From $438 to $443

4) From 256 ft to 140 ft

5) From 6469 ft to 7488 ft

6) From 36 inches to 90 inches

7) From 54 ft to 104 ft

8) From 84 miles to 24 miles

9) The population of a place in a particular year increased by 15%. Next year it decreased by 15%. Find the net increase or decrease per cent in the initial population.

10) The salary of a doctor is increased by 40%. By what per cent should the new salary be reduced in order to restore the original salary?

4-8 Markup, Discount, and Tax

Find the selling price of each item.

1) Original price of a microphone: $49.99, discount: 5%, tax: 5%

2) Cost of a pen: $1.95, markup: 70%, discount: 40%, tax: 5%

3) Cost of a puppy: $349.99, markup: 41%, discount: 23%

4) Cost of a shirt: $14.95, markup: 25%, discount: 45%

5) Cost of an oil change: $21.95, markup: 95%

6) Cost of computer: $1,850.00, markup: 75%

Arithmetic and Pre-Algebra Workbook

Answers of Worksheets – Chapter 4

4–1 Converting Between Percent, Fractions, and Decimals

1) 0.5
2) 0.38
3) 0.15
4) 0.8
5) 0.07
6) 0.35
7) 0.9
8) 0.2
9) 0.07
10) 50%
11) 90%
12) 0.2%
13) 52.4%
14) 10%
15) 3%
16) 363%
17) 0.8%
18) 478%

4–2 Table of Common Percent

Fraction	Decimal	Percent
$\frac{1}{25}$	0.04	4%
$\frac{1}{2}$	0.5	50%
$\frac{1}{4}$	0.25	25%
$\frac{1}{5}$	0.2	20%
$\frac{6}{10}$	0.6	60%
$\frac{5}{8}$	0.625	62.5 %
$\frac{2}{5}$	0.4	40%
$\frac{7}{100}$	0.07	7%
$\frac{7}{16}$	0.4375	43.75%
$\frac{5}{8}$	0.625	625 %

www.EffortlessMath.com

Arithmetic and Pre-Algebra Workbook

$\frac{7}{10}$	0.7	70%
$\frac{30}{100}$	0.3	30%
$\frac{4}{8}$	0.8	80%
$\frac{3}{4}$	0.75	75%

4–3 Percentage Calculations

1) 12.5
2) 12
3) 10.2
4) 31.5
5) 0
6) 17.6
7) 5.2
8) 42.12
9) 40
10) 2
11) 16
12) 0
13) 14
14) 33
15) 8
16) 176
17) 70
18) 72
19) 67%
20) 70%
21) 58%
22) 50%

4–4 Find What Percentage a Number Is of Another

1) 45 is what percent of 90? 50 %
2) 15 is what percent of 75? 20 %
3) 20 is what percent of 400? 5 %
4) 18 is what percent of 90? 20 %
5) 3 is what percent of 15? 20 %
6) 8 is what percent of 80? 10 %
7) 11 is what percent of 55? 20 %
8) 9 is what percent of 90? 10 %
9) 2.5 is what percent of 10? 25 %
10) 5 is what percent of 25? 20 %
11) 60 is what percent of 20? 300 %
12) 12 is what percent of 48? 25 %
13) 14 is what percent of 28? 50 %
14) 8.2 is what percent of 32.8? 25 %
15) 1200 is what percent of 4,800? 25%
16) 4,000 is what percent of 20,000? 20 %
17) 45 is what percent of 900? 5 %
18) 10 is what percent of 200? 5 %

19) 15 is what percent of 60? 25 %   20) 1.2 is what percent of 24? 5 %

4–5 Find a Percentage of a Given Number

1) 45	7) 6	13) 33.6	19) 0.9
2) 20	8) 3.5	14) 10.8	20) 20
3) 0	9) 8	15) 11.2	21) 8
4) 64	10) 6	16) 16	
5) 24	11) 50	17) 2.4	
6) 30	12) 30.6	18) 26.6	

4–6 Percent Problems

1) 15	5) 47.6%	9) 1.9%	13) 49 hours
2) 104.3	6) 100	10) 8.3 m	14) 62.7%
3) 38.34	7) 515.4	11) 117 inches	15) Mathematics
4) 23.44%	8) 31.98	12) 45.7 inches	16) 600

4–7 Percent of Increase and Decrease

1) 156.25% increase
2) 70% decrease
3) 1.142% increase
4) 45.31% decrease
5) 15.75% increase
6) 150% increase
7) 92.6% increase
8) 71.43% decrease
9) 2.25% decrease
10) $28\frac{4}{7}$ %

4-8 Markup, Discount, and Tax

1) $49.87
2) $2.09
3) $379.98
4) $10.28
5) $42.80
6) $3,237.50

Chapter 5: Algebraic Expressions

5–1 Expressions and Variables

5–2 Simplifying Variable Expressions

5–3 The Distributive Property

5–4 Translate Phrases into an Algebraic Statement

5–5 Evaluating One Variable

5–6 Evaluating Two Variables

5–7 Combining like Terms

5–8 Simplifying Polynomial Expressions

5-1 Expressions and Variables

Simplify each expression.

1) $x + 5x$,

   use $x = 5$

2) $8(-3x + 9) + 6$,

   use $x = 6$

3) $10x - 2x + 6 - 5$,

   use $x = 5$

4) $2x - 3x - 9$,

   use $x = 7$

5) $(-6)(-2x - 4y)$,

   use $x = 1, y = 3$

6) $8x + 2 + 4y$,

   use $x = 9, y = 2$

7) $(-6)(-8x - 9y)$,

   use $x = 5, y = 5$

8) $6x + 5y$,

   use $x = 7, y = 4$

Simplify each expression.

9) $5(-4 + 2x)$

10) $-3 - 5x - 6x + 9$

11) $6x - 3x - 8 + 10$

12) $(-8)(6x - 4) + 12$

13) $9(7x + 4) + 6x$

14) $(-9)(-5x + 2)$

5-2 Simplifying Variable Expressions

Simplify each expression.

1) $-2 - x^2 - 6x^2$

2) $3 + 10x^2 + 2$

3) $8x^2 + 6x + 7x^2$

4) $5x^2 - 12x^2 + 8x$

5) $2x^2 - 2x - x$

6) $(-6)(8x - 4)$

7) $4x + 6(2 - 5x)$

8) $10x + 8(10x - 6)$

9) $9(-2x - 6) - 5$

10) $3(x + 9)$

11) $7x + 3 - 3x$

12) $2.5x^2 \times (-8x)$

Simplify.

13) $-2(4 - 6x) - 3x$, $x = 1$

14) $2x + 8x$, $x = 2$

15) $9 - 2x + 5x + 2$, $x = 5$

16) $5(3x + 7)$, $x = 3$

17) $2(3 - 2x) - 4$, $x = 6$

18) $5x + 3x - 8$, $x = 3$

19) $x - 7x$, $x = 8$

20) $5(-2 - 9x)$, $x = 4$

5-3 The Distributive Property

Use the distributive property to simply each expression.

1) $-(-2-5x)$

2) $(-6x+2)(-1)$

3) $(-5)(x-2)$

4) $-(7-3x)$

5) $8(8+2x)$

6) $2(12+2x)$

7) $(-6x+8)4$

8) $(3-6x)(-7)$

9) $(-12)(2x+1)$

10) $(8-2x)9$

11) $(-2x)(-1+9x)-4x(4+5x)$

12) $3(-5x-3)+4(6-3x)$

13) $(-2)(x+4)-(2+3x)$

14) $(-4)(3x-2)+6(x+1)$

15) $(-5)(4x-1)+4(x+2)$

16) $(-3)(x+4)-(2+3x)$

5-4 Translate Phrases into an Algebraic Statement

Write an algebraic expression for each phrase.

1) A number increased by forty-two.

2) The sum of fifteen and a number

3) The difference between fifty-six and a number.

4) The quotient of thirty and a number.

5) Twice a number decreased by 25.

6) Four times the sum of a number and – 12.

7) A number divided by – 20.

8) The quotient of 60 and the product of a number and – 5.

9) Ten subtracted from a number.

10) The difference of six and a number.

5-5 Evaluating One Variable

Simplify each algebraic expression.

1) $9 - x$, $x = 3$

2) $x + 2$, $x = 5$

3) $3x + 7$, $x = 6$

4) $x + (-5)$, $x = -2$

5) $3x + 6$, $x = 4$

6) $4x + 6$, $x = -1$

7) $10 + 2x - 6$, $x = 3$

8) $10 - 3x$, $x = 8$

9) $\frac{20}{x} - 3$, $x = 5$

10) $(-3) + \frac{x}{4} + 2x$, $x = 16$

11) $(-2) + \frac{x}{7}$, $x = 21$

12) $(-\frac{14}{x}) - 9 + 4x$, $x = 2$

13) $(-\frac{6}{x}) - 9 + 2x$, $x = 3$

14) $(-2) + \frac{x}{8}$, $x = 16$

15) $8(5x - 12)$, $x = -2$

5-6 Evaluating Two Variables

Simplify each algebraic expression.

1) $2x + 4y - 3 + 2$,

    $x = 5, y = 3$

2) $(-\dfrac{12}{x}) + 1 + 5y$,

    $x = 6, y = 8$

3) $(-4)(-2a - 2b)$,

    $a = 5, b = 3$

4) $10 + 3x + 7 - 2y$,

    $x = 7, y = 6$

5) $9x + 2 - 4y$,

    $x = 7, y = 5$

6) $6 + 3(-2x - 3y)$,

    $x = 9, y = 7$

7) $12x + y$,

    $x = 4, y = 8$

8) $x \times 4 \div y$,

    $x = 3, y = 2$

9) $2x + 14 + 4y$,

    $x = 6, y = 8$

10) $4a - (5 - b)$,

    $a = 4, b = 6$

5-7 Combining like Terms

Simplify each expression.

1) $5 + 2x - 8$

2) $(-2x + 6)\,2$

3) $7 + 3x + 6x - 4$

4) $(-4) - (3)(5x + 8)$

5) $9x - 7x - 5$

6) $x - 12x$

7) $7(3x + 6) + 2x$

8) $(-11x) - 10x$

9) $3x - 12 - 5x$

10) $13 + 4x - 5$

11) $(-22x) + 8x$

12) $2(4 + 3x) - 7x$

13) $(-4x) - (6 - 14x)$

14) $5(6x - 1) + 12x$

15) $22x + 6 + 2x$

16) $(-13x) - 14x$

17) $(-6x) - 9 + 15x$

18) $(-6x) + 7x$

19) $(-5x) + 12 + 7x$

20) $(-3x) - 9 + 15x$

21) $20x - 19x$

5-8 Simplifying Polynomial Expressions

Simplify each polynomial.

1) $4x^5 - 5x^6 + 15x^5 - 12x^6 + 3x^6$

2) $(-3x^5 + 12 - 4x) + (8x^4 + 5x + 5x^5)$

3) $10x^2 - 5x^4 + 14x^3 - 20x^4 + 15x^3 - 8x^4$

4) $-6x^2 + 5x^2 - 7x^3 + 12 + 22$

5) $12x^5 - 5x^3 + 8x^2 - 8x^5$

6) $5x^3 + 1 + x^2 - 2x - 10x$

7) $14x^2 - 6x^3 - 2x(4x^2 + 2x)$

8) $(4x^4 - 2x) - (4x - 2x^4)$

9) $(3x^2 + 1) - (4 + 2x^2)$

10) $(2x + 2) - (7x + 6)$

11) $(12x^3 + 4x^4) - (2x^4 - 6x^3)$

12) $(12 + 3x^3) + (6x^3 + 6)$

13) $(5x^2 - 3) + (2x^2 - 3x^3)$

14) $(23x^3 - 12x^2) - (2x^2 - 9x^3)$

15) $(4x - 3x^3) - (3x^3 + 4x)$

Answers of Worksheets – Chapter 5

5–1 Expressions and Variables

1) 30
2) −66
3) 41
4) −16
5) 84
6) 82
7) 510
8) 62
9) 10x − 20
10) 6 − 11x
11) 3x + 2
12) 44 − 48x
13) 69x + 36
14) 45x − 18

5–2 Simplifying Variable Expressions

1) $-7x^2 - 2$
2) $10x^2 + 5$
3) $15x^2 + 6x$
4) $-7x^2 + 8x$
5) $2x^2 - 3x$
6) $-48x + 24$
7) $-26x + 12$
8) $90x - 48$
9) $-18x - 59$
10) $3x + 27$
11) $4x + 3$
12) $-20x^3$
13) 1
14) 20
15) 26
16) 80
17) −22
18) 16
19) −48
20) −190

5–3 The Distributive Property

1) 5x + 2
2) 6x − 2
3) −5x + 10
4) 3x − 7
5) 16x + 64
6) 4x + 24
7) −24x + 32
8) 42x − 21
9) −24x − 12
10) −18x + 72
11) $-38x^2 - 14x$
12) −27x + 15
13) −5x − 10
14) −6x + 14
15) −16x + 13
16) −6x − 14

Arithmetic and Pre-Algebra Workbook

5–4 Translate Phrases into an Algebraic Statement

1) x + 42
2) 15 + x
3) 56 – x
4) 30/x
5) 2x – 25
6) 4(x + (–12))
7) $\frac{x}{-20}$
8) $\frac{60}{-5x}$
9) x – 10
10) 6 – x

5–5 Evaluating One Variable

1) 6
2) 7
3) 25
4) –7
5) 18
6) 2
7) 10
8) –14
9) 1
10) 33
11) 1
12) –8
13) –5
14) 0
15) –176

5–6 Evaluating Two Variables

1) 21
2) 39
3) 64
4) 26
5) 45
6) –111
7) 56
8) 6
9) 58
10) 17

5–7 Combining like Terms

1) 2x – 3
2) –4x + 12
3) 9x + 3
4) –15x – 28
5) 2x – 5
6) –11x
7) 23x + 42
8) –21x
9) –2x – 12
10) 4x + 8
11) –14x
12) – x + 8
13) 10x – 6
14) 42x – 5
15) 24x + 6
16) –27x
17) 9x – 9
18) x
19) 2x + 12
20) 12x – 9
21) x

www.EffortlessMath.com

5–8 Simplifying Polynomial Expressions

1) $-14x^6 + 19x^5$

2) $2x^5 + 8x^4 + x + 12$

3) $-33x^4 + 29x^3 + 10x^2$

4) $-7x^3 - x^2 + 34$

5) $4x^5 - 5x^3 + 8x^2$

6) $5x^3 + x^2 - 12x + 1$

7) $-14x^3 + 10x^2$

8) $6x^4 - 6x$

9) $x^2 - 3$

10) $-5x - 4$

11) $2x^4 + 18x^3$

12) $9x^3 + 18$

13) $-3x^3 + 7x^2 - 3$

14) $32x^3 - 14x^2$

15) $-6x^3$

Chapter 6: Equations

6–1 One–Step Equations

6–2 One–Step Equation Word Problems

6–3 Two–Step Equations

6–4 Two–Step Equation Word Problems

6–5 Multi–Step Equations

6-1 One-Step Equations

Solve each equation.

1) $x + 3 = 17$

2) $22 = (-8) + x$

3) $3x = (-30)$

4) $(-36) = (-6x)$

5) $(-6) = 4 + x$

6) $2 + x = (-2)$

7) $20x = (-220)$

8) $18 = x + 5$

9) $(-23) + x = (-19)$

10) $5x = (-45)$

11) $x - 12 = (-25)$

12) $x - 3 = (-12)$

13) $(-35) = x - 27$

14) $8 = 2x$

15) $(-6x) = 36$

16) $(-55) = (-5x)$

17) $x - 30 = 20$

18) $8x = 32$

19) $36 = (-4x)$

20) $4x = 68$

21) $30x = 300$

6-2 One-Step Equation Word Problems

Solve.

1) How many boxes of envelopes can you buy with $18 if one box costs $3?

2) After paying $6.25 for a salad, Ella has $45.56. How much money did she have before buying the salad?

3) How many packages of diapers can you buy with $50 if one package costs $5?

4) Last week James ran 20 miles more than Michael. James ran 56 miles. How many miles did Michael run?

5) Last Friday Jacob had $32.52. Over the weekend he received some money for cleaning the attic. He now has $44. How much money did he receive?

6) After paying $10.12 for a sandwich, Amelia has $35.50. How much money did she have before buying the sandwich?

6-3 Two-Step Equations

Solve each equation.

1) $5(8 + x) = 20$

2) $(-7)(x - 9) = 42$

3) $(-12)(2x - 3) = (-12)$

4) $6(1 + x) = 12$

5) $12(2x + 4) = 60$

6) $7(3x + 2) = 42$

7) $8(14 + 2x) = (-34)$

8) $(-15)(2x - 4) = 48$

9) $3(x + 5) = 12$

10) $\dfrac{3x - 12}{6} = 4$

11) $(-12) = \dfrac{x + 15}{6}$

12) $110 = (-5)(2x - 6)$

13) $\dfrac{x}{8} - 12 = 4$

14) $20 = 12 + \dfrac{x}{4}$

15) $\dfrac{-24 + x}{6} = (-12)$

16) $(-4)(5 + 2x) = (-100)$

17) $(-12x) + 20 = 32$

18) $\dfrac{-2 + 6x}{4} = (-8)$

19) $\dfrac{x + 6}{5} = (-5)$

20) $(-9) + \dfrac{x}{4} = (-15)$

6-4 Two-Step Equation Word Problems

Solve.

1) The sum of three consecutive even numbers is 48. What is the smallest of these numbers?

2) How old am I if 400 reduced by 2 times my age is 244?

3) For a field trip, 4 students rode in cars and the rest filled nine buses. How many students were in each bus if 472 students were on the trip?

4) The sum of three consecutive numbers is 72. What is the smallest of these numbers?

5) 331 students went on a field trip. Six buses were filled, and 7 students traveled in cars. How many students were in each bus?

6) You bought a magazine for $5 and four erasers. You spent a total of $25. How much did each eraser cost?

6-5 Multi-Step Equations

Solve each equation.

1) $-(2 - 2x) = 10$

2) $-12 = -(2x + 8)$

3) $3x + 15 = (-2x) + 5$

4) $-28 = (-2x) - 12x$

5) $2(1 + 2x) + 2x = -118$

6) $3x - 18 = 22 + x - 3 + x$

7) $12 - 2x = (-32) - x + x$

8) $7 - 3x - 3x = 3 - 3x$

9) $6 + 10x + 3x = (-30) + 4x$

10) $(-3x) - 8(-1 + 5x) = 352$

11) $24 = (-4x) - 8 + 8$

12) $9 = 2x - 7 + 6x$

13) $6(1 + 6x) = 294$

14) $-10 = (-4x) - 6x$

15) $4x - 2 = (-7) + 5x$

16) $5x - 14 = 8x + 4$

17) $40 = -(4x - 8)$

18) $(-18) - 6x = 6(1 + 3x)$

19) $x - 5 = -2(6 + 3x)$

20) $6 = 1 - 2x + 5$

Arithmetic and Pre-Algebra Workbook

Answers of Worksheets – Chapter 6

6–1 One–Step Equations

1) 14
2) 30
3) −10
4) 6
5) −10
6) −4
7) −11
8) 13
9) 4
10) −9
11) −13
12) −9
13) −8
14) 4
15) −6
16) 11
17) 50
18) 4
19) −9
20) 17
21) 10

6–2 One–Step Equation Word Problems

1) 6
2) $51.81
3) 10
4) 36
5) 11.48
6) 45.62

6–3 Two–Step Equations

1) −4
2) 3
3) 2
4) 1
5) 0.5
6) $\frac{4}{3}$
7) $-\frac{73}{8}$
8) $\frac{2}{5}$
9) −1
10) 12
11) −87
12) −8
13) 128
14) 32
15) −48
16) 10
17) −1
18) −5
19) −31
20) −24

6–4 Two–Step Equation Word Problems

1) 14
2) 78
3) 52
4) 23
5) 54
6) $4

6–5 Multi–Step Equations

1) 6
2) 2
3) −2
4) 2
5) −20
6) 37
7) 22
8) $\frac{4}{3}$
9) −4
10) −8
11) −6
12) 2
13) 8
14) 1
15) 5
16) −6
17) −8
18) −1
19) −1
20) 0

Chapter 7: Systems of Equations

7–1 Solving Systems of Equations by Graphing

7–2 Solving Systems of Equations by Substitution

7–3 Solving Systems of Equations by Elimination

7–4 Systems of Equations Word Problems

7-1 Solving Systems of Equations by Graphing

Solve each system of equations by graphing.

1) $y = -4x - 2$
   $y = -2x + 1$

2) $y = -8x - 4$
   $y = 2$

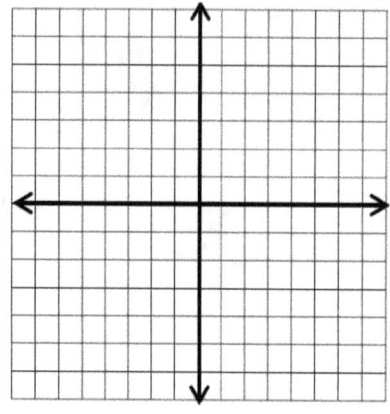
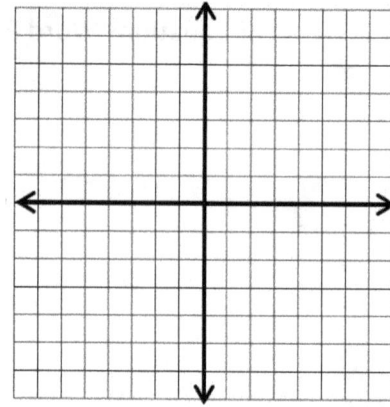

4) $y = 9x - 5$
   $y = -6x + 4$

5) $y = \frac{1}{2}x + 8$
   $y = 8x - 2$

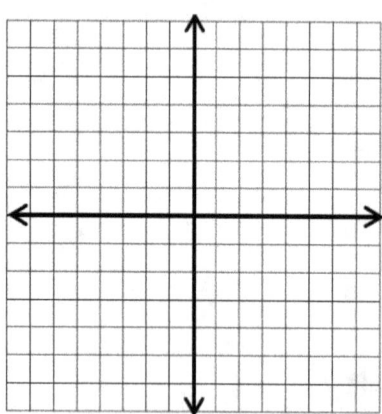
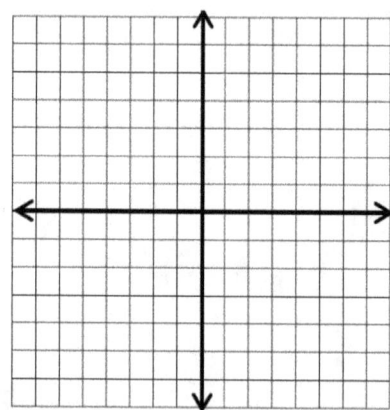

7-2 Solving Systems of Equations by Substitution

Solve each system of equation by substitution.

1) $-x - y = -13$
   $-2x + 2y = 10$

2) $-2x + 2y = 4$
   $-2x + y = 3$

3) $-10x + 2y = -6$
   $6x - 16y = 48$

4) $y = -8$
   $x - 12y = 72$

5) $2y = -6x + 10$
   $10x - 8y = -6$

6) $3x - 9y = -3$
   $3y = 3x - 3$

7) $-4x + 12y = 12$
   $-14x + 16y = -10$

8) $-10x - 16y = 34$
   $4x - 14y = -34$

7-3 Solving Systems of Equations by Elimination

Solve each system of equation by elimination.

1) $x - y = -12$
   $-5x + 3y = 6$

2) $-3x - 4y = 5$
   $x - 2y = 5$

3) $5x - 14y = 22$
   $-6x + 7y = 3$

4) $10x - 14y = -4$
   $-10x - 20y = -30$

5) $32x + 14y = 52$
   $16x - 4y = -40$

6) $2x - 8y = -6$
   $8x + 2y = 10$

7) $-4x + 4y = -4$
   $4x + 2y = 10$

8) $4x + 6y = 10$
   $8x + 12y = -20$

9) $20x - 18y = -12$
   $18x - 8y = 22$

10) $8x + 10y = 52$
    $8x + 6y = 44$

7-4 Systems of Equations Word Problems

Solve.

1) A farmhouse shelters 10 animals, some are pigs and some are ducks. Altogether there are 36 legs. How many of each animal are there?

2) A class of 195 students went on a field trip. They took vehicles, some cars and some buses. Find the number of cars and the number of buses they took if each car holds 5 students and each bus hold 45 students.

3) The difference of two numbers is 6. Their sum is 14. Find the numbers.

4) The sum of the digits of a certain two–digit number is 7. Reversing its increasing the number by 9. What is the number?

5) The sum of two numbers is 66. Their difference is 18. Find the numbers.

Answers of Worksheets – Chapter 7

7–1 Solving Systems of Equations by Graphing

1)
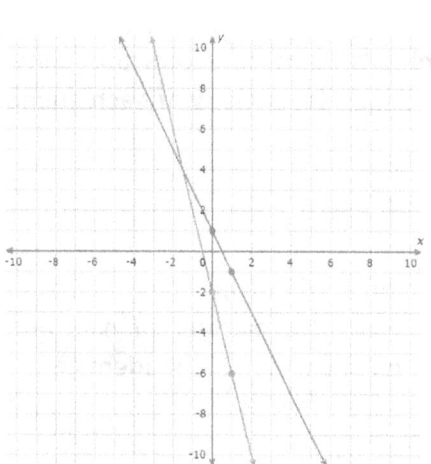

2)
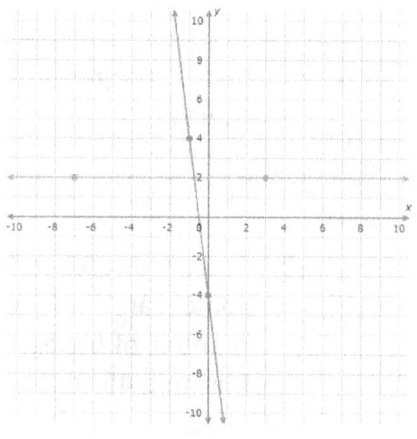

3)
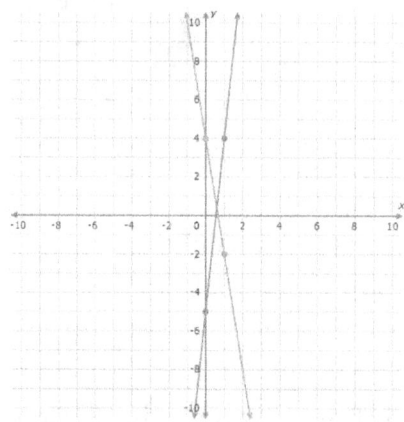

4)
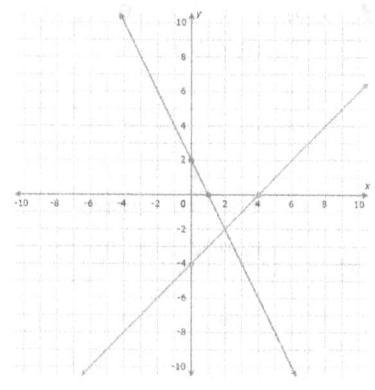

7–2 Solving Systems of Equations by Substitution

1) (4, 9)
2) (−1, 1)
3) (0, −3)
4) (−24, −8)
5) (1, 2)
6) (2, 1)

Arithmetic and Pre-Algebra Workbook

7) (3, 2) 8) (−5, 1)

7–3 Solving Systems of Equations by Elimination

1) (15, 27) 5) (−1, 6) 9) (3, 4)

2) (1, −2) 6) (1, 1) 10) (4, 2)

3) (−4, −3) 7) (2, 1)

4) (1, 1) 8) No solution

7–4 Systems of Equations Word Problems

1) There are 8 pigs and 2 ducks.

2) There are 3 cars and 4 buses.

3) 10 and 4

4) 34

5) 24 and 42

Chapter 8: Inequalities

8–1 Graphing Single–Variable Inequalities

8–2 One–Step Inequalities

8–3 Two–Step Inequalities

8–4 Multi–Step Inequalities

8-1 Graphing Single-Variable Inequalities

Draw a graph for each inequality.

1) $-2 > x$

2) $5 \leq -x$

3) $x > 7$

4) $-x > 1.5$

Arithmetic and Pre-Algebra Workbook

8-2 One-Step Inequalities

Solve each inequality and graph it.

1) $x + 9 \geq 11$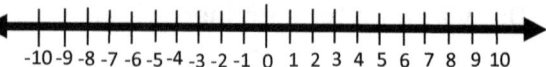

2) $x - 4 \leq 2$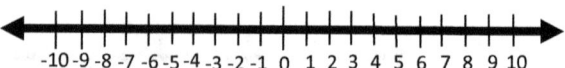

3) $6x \geq 36$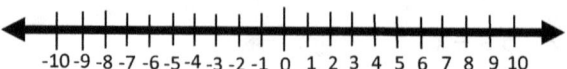

4) $7 + x < 16$

5) $x + 8 \leq 1$

6) $3x > 12$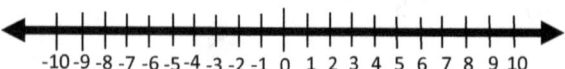

7) $3x < 24$

8-3 Two-Step Inequalities

Solve each inequality and graph it.

1) $3x - 4 \leq 5$

2) $4x - 19 < 19$

3) $3x + 6 \geq 12$

4) $6x - 5 \geq 19$

5) $2x - 3 < 21$

6) $3 + 4x < 19$

8-4 Multi-Step Inequalities

Solve each inequality and graph it.

1) $\dfrac{7x+1}{3} \geq 5$

2) $\dfrac{9x}{7} - x < 2$

3) $\dfrac{4x+8}{2} \leq 12$

4) $\dfrac{3x-8}{7} > 1$

5) $-3(x-7) > 21$

6) $4 + \dfrac{x}{3} < 7$

Arithmetic and Pre-Algebra Workbook

Answers of Worksheets – Chapter 8

8–1 Graphing Single–Variable Inequalities

1) $-2 > x$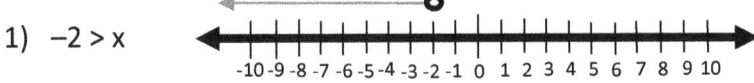

2) $x \leq -5$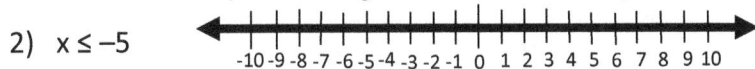

3) $x > 7$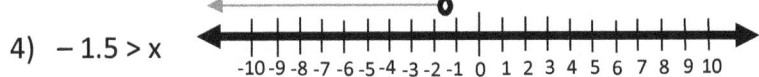

4) $-1.5 > x$

8–2 One–Step Inequalities

1) $x + 9 \geq 11$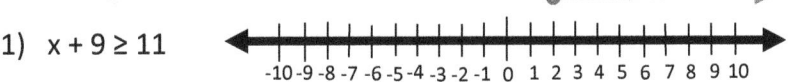

2) $x - 4 \leq 2$

3) $6x \geq 36$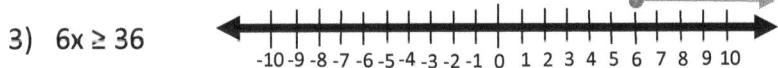

4) $7 + x < 16$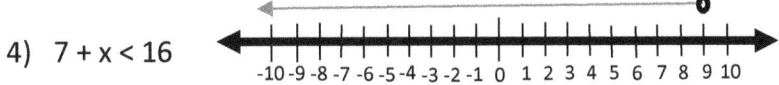

5) $x + 8 \leq 1$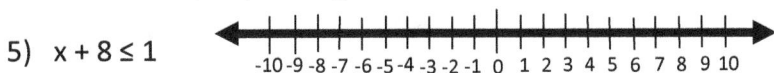

6) $3x > 12$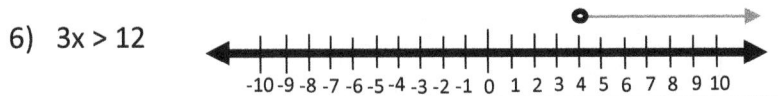

www.EffortlessMath.com

7) $3x < 24$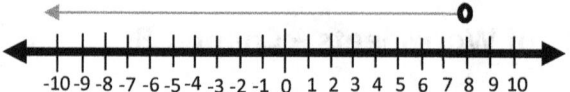

8–3 Two–Step Inequalities

1) $3x - 4 \leq 5$

2) $4x + 19 < 19$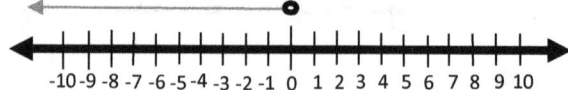

3) $3x + 6 \geq 12$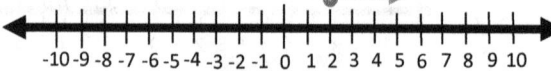

4) $6x - 5 \geq 19$

5) $2x - 3 < 11$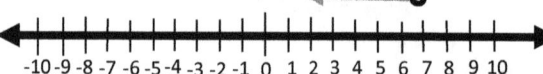

6) $3 + 4x < 19$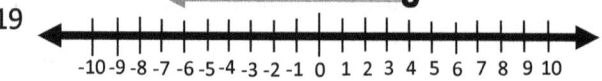

8–4 Multi–Step Inequalities

1) $\dfrac{7x+1}{3} \geq 5$

2) $\dfrac{9x}{7} - x < 2$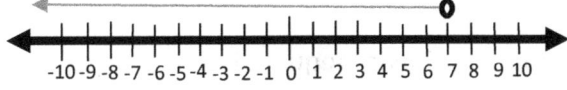

3) $\dfrac{4x+8}{2} \leq 12$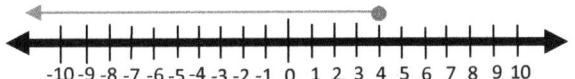

4) $\dfrac{3x-8}{7} > 1$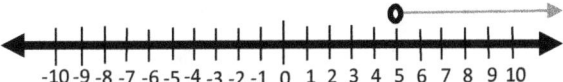

5) $-3(x-7) > 21$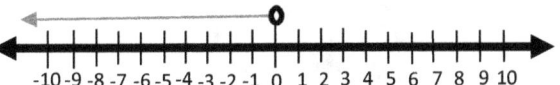

6) $4 + \dfrac{x}{3} < 7$

Chapter 9: Linear Functions

9–1 Finding Slope

9–2 Graphing Lines Using Slope–Intercept Form

9–3 Graphing Lines Using Standard Form

9–4 Writing Linear Equations

9–5 Graphing Linear Inequalities

9–6 Finding Midpoint

9–7 Finding Distance of Two Points

9-1 Finding Slope

Find the slope of the line through each pair of points.

1) $(2, -10), (3, 6)$

2) $(4, -6), (-3, -8)$

3) $(7, -12), (5, 10)$

4) $(19, 3), (20, 3)$

5) $(15, 8), (-17, 9)$

6) $(6, -12), (15, -3)$

7) $(3, 1), (7, -5)$

8) $(3, -2), (-7, 8)$

9) $(15, -3), (-9, 5)$

10) $(-4, 7), (-6, -4)$

11) $(6, -8), (-11, -7)$

12) $(-6, 13), (17, -9)$

13) $(-10, -2), (-6, -5)$

14) $(4, 5), (-4, 10)$

15) $(-3, 1), (-17, 2)$

16) $(7, 0), (-13, -11)$

17) $(17, -13), (17, 8)$

18) $(12, 2), (-7, 5)$

www.EffortlessMath.com

Arithmetic and Pre-Algebra Workbook

9-2 Graphing Lines Using Slope-Intercept Form

Sketch the graph of each line.

1) $y = \dfrac{2}{3}x - 3$

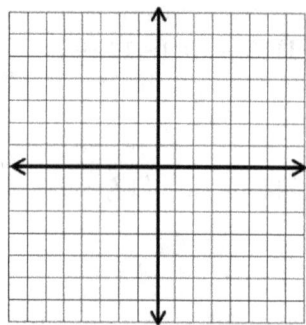

2) $y = -\dfrac{4}{5}x - 5$

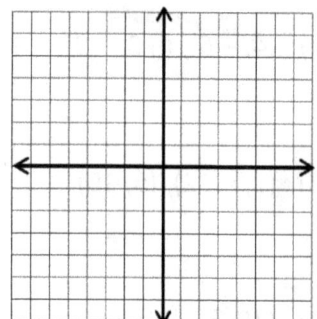

4) $y = \dfrac{1}{5}x - 4$

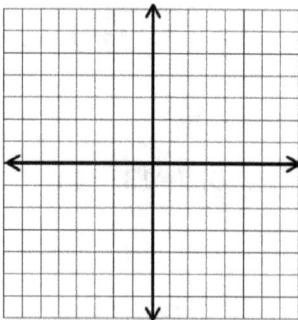

5) $y = 2x$

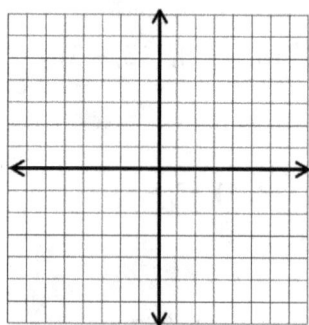

Arithmetic and Pre-Algebra Workbook

9-3 Graphing Lines Using Standard Form

Sketch the graph of each line.

1) $x + 4y = 12$

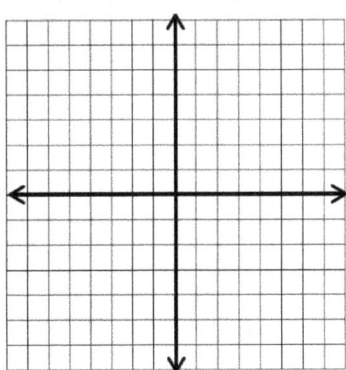

2) $2y = -2$

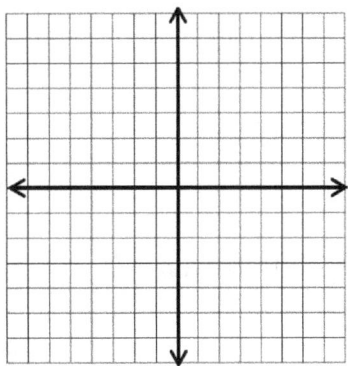

4) $2x - y = 4$

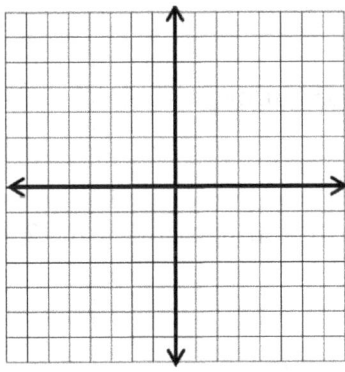

5) $x + y = 2$

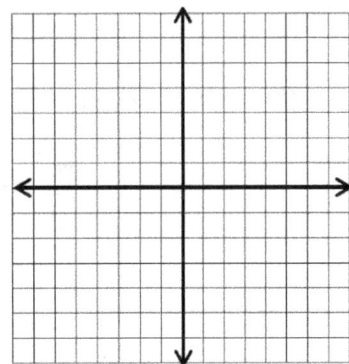

www.EffortlessMath.com

9-4 Writing Linear Equations

Write the slope–intercept form of the equation of the line through the given points.

1) through: (− 4, − 2), (− 3, 5)

2) through: (5, 4), (− 4, 3)

3) through: (0, − 2), (− 5, 3)

4) through: (− 4, − 2), (− 3, 5)

5) through: (0, 3), (− 4, − 1)

6) through: (0, 2), (1, − 3)

7) through: (0, − 5), (4, 3)

8) through: (− 1, 4), (0, 4)

9) through: (2, − 3), (3, − 5)

10) through: (2, 5), (− 1, − 4)

11) through: (1, − 3), (− 3, 1)

12) through: (3, 3), (1, − 5)

13) through: (4, 4), (3, − 5)

14) through: (0, 3), (1, 1)

15) through: (5, 5), (2, − 3)

16) through: (− 2, − 2), (2, − 5)

17) through: (− 3, − 2), (1, − 1)

18) through: (1, 5), (4, 1)

9-5 Graphing Linear Inequalities

Sketch the graph of each linear inequality.

1) $y < -4x + 2$

2) $2x + y < -4$

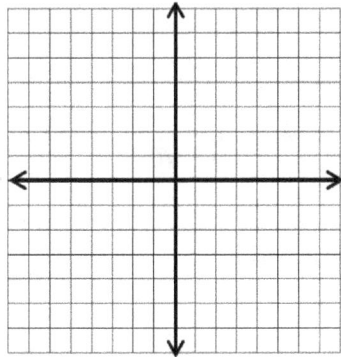 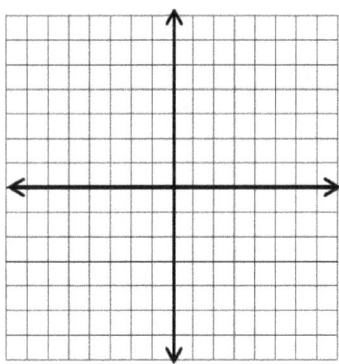

4) $x - 3y < -5$

5) $6x - 2y \geq 8$

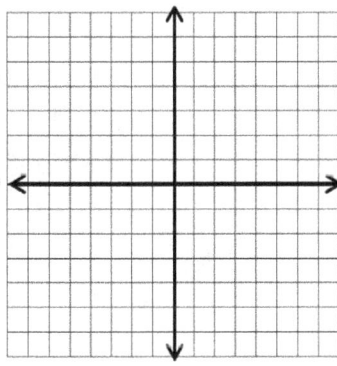 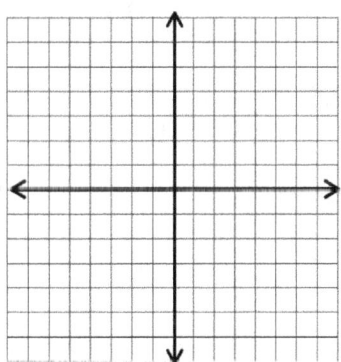

9-6 Finding Midpoint

Find the midpoint of the line segment with the given endpoints.

1) (3, 9), (− 1, 6)

2) (2, − 2), (3, − 5)

3) (− 2, 6), (− 3, − 2)

4) (0, 2), (− 2, − 6)

5) (7, 4), (9, − 1)

6) (4, − 5), (0, 8)

7) (1, − 2), (1, − 6)

8) (− 2, − 3), (3, − 6)

9) (7, 0), (− 7, 5)

10) (− 2, 6), (− 3, − 2)

11) (− 1, 1), (5, − 5)

12) (2.3, − 1.3), (− 2.2, − 0.5)

13) (4.1, 6.32), (4, 5.6)

14) (2, − 1), (− 6, 0)

15) (− 4, 4), (5, − 1)

16) (− 2, − 3), (− 6, 5)

17) ($\frac{1}{2}$, 1), (2, 4)

18) (− 2.9, − 2.958), (8.6, 5)

9-7 Finding Distance of Two Points

Find the distance between each pair of points.

1) $(-1, 2), (-1, -7)$

2) $(6, 4), (-1, 3)$

3) $(-8, -5), (-6, 1)$

4) $(-6, -10), (-2, -10)$

5) $(4, -6), (-3, 4)$

6) $(-6, -7), (-2, -8)$

7) $(5, 4), (8, 2)$

8) $(8, 4), (3, -7)$

9) $(1, 3), (5, 7)$

10) $(4, 2), (-7, 1)$

11) $(-3, -4), (-7, -2)$

12) $(-7, -2), (6, 9)$

13) $(10, 0), (0, 4)$

14) $(-3, 2), (5, 0)$

15) $(-5, 6), (8, -4)$

16) $(3, -5), (-8, -4)$

17) $(0, 8), (4, 10)$

18) $(6, 4), (-5, -1)$

Answers of Worksheets – Chapter 9

9–1 Finding Slope

1) 16
2) $\frac{2}{7}$
3) -11
4) 0
5) $-\frac{1}{32}$
6) 1
7) $-\frac{3}{2}$
8) -1
9) $-\frac{1}{3}$
10) $\frac{11}{2}$
11) $-\frac{1}{17}$
12) $-\frac{22}{23}$
13) $-\frac{3}{4}$
14) $-\frac{5}{8}$
15) $-\frac{1}{14}$
16) $\frac{11}{20}$
17) Undefined
18) $-\frac{3}{19}$
19) $-\frac{3}{19}$

9–2 Graphing Lines Using Slope–Intercept Form

1) 2)

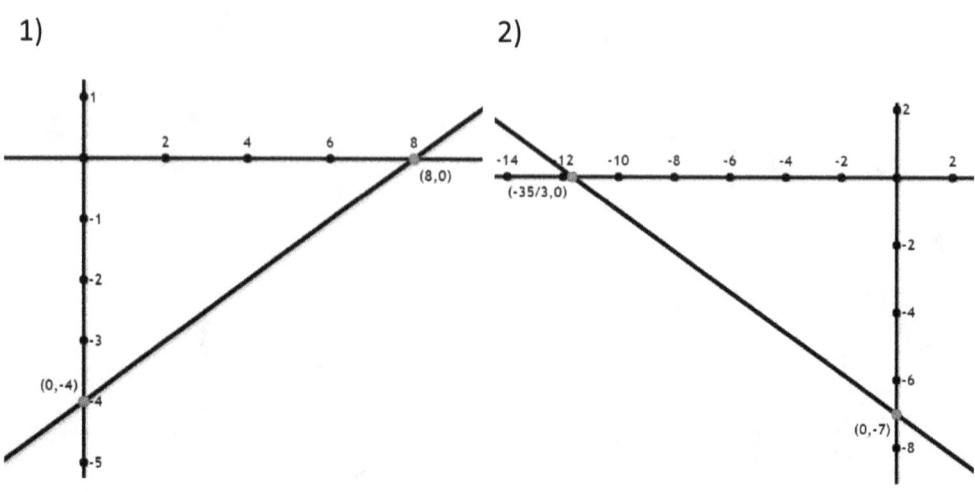

3)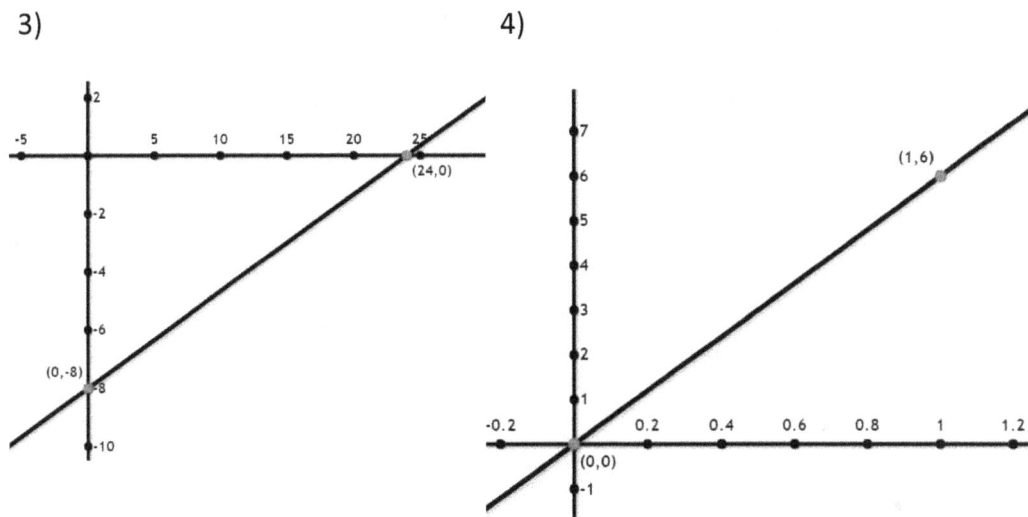

4)

9–3 Graphing Lines Using Standard Form

1)

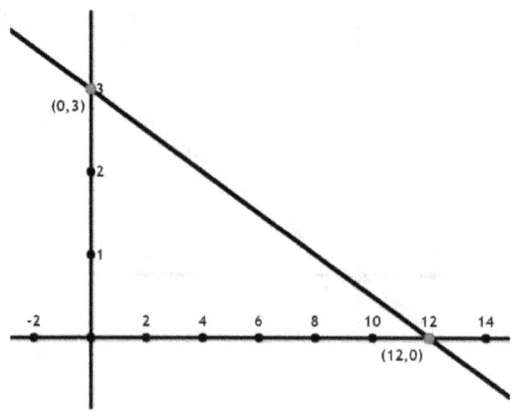

2)

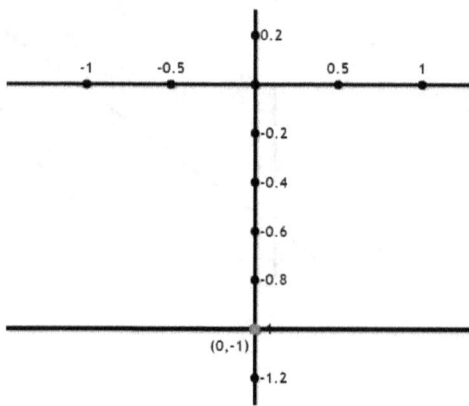

3)

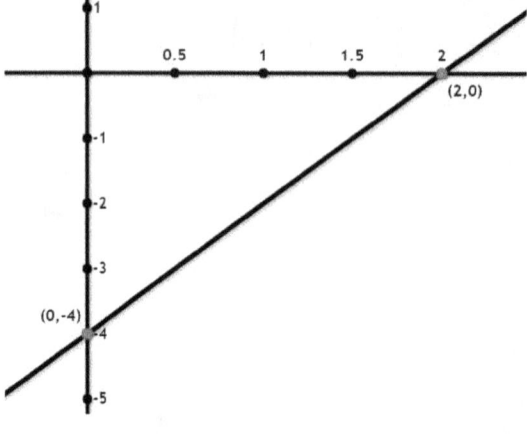

4)

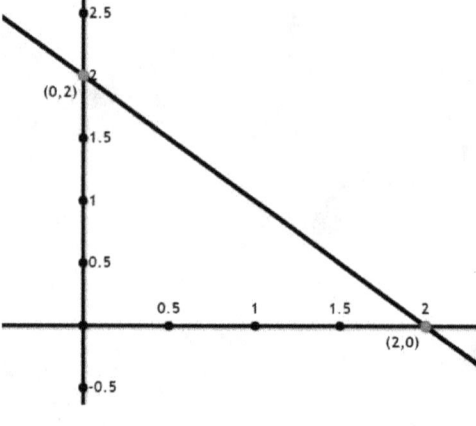

9–4 Writing Linear Equations

1) $y = 7x + 26$
2) $y = \frac{1}{4}x + 4$
3) $y = -x - 2$
4) $y = 7x + 26$
5) $y = x + 3$
6) $y = -5x + 2$
7) $y = 2x - 5$
8) $y = 4$
9) $y = -2x + 1$
10) $y = 3x - 1$
11) $y = -x - 2$
12) $y = 4x - 9$
13) $y = 9x - 32$
14) $y = -2x + 3$
15) $y = \frac{8}{3}x - \frac{25}{3}$
16) $y = -\frac{3}{4}x - \frac{7}{2}$
17) $y = -\frac{3}{4}x - \frac{1}{4}$
18) $y = -\frac{4}{3}x + \frac{19}{3}$

Arithmetic and Pre-Algebra Workbook

9–5 Graphing Linear Inequalities

1)

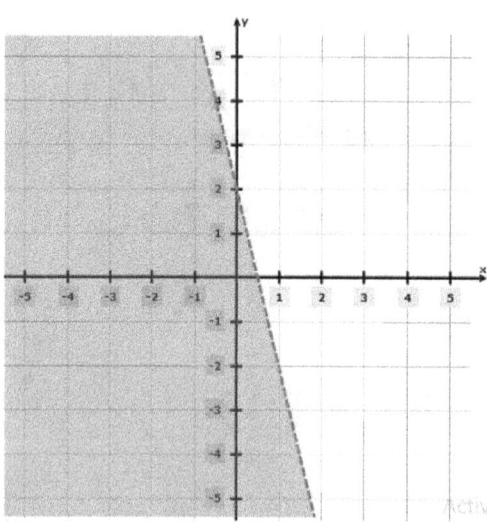

2)

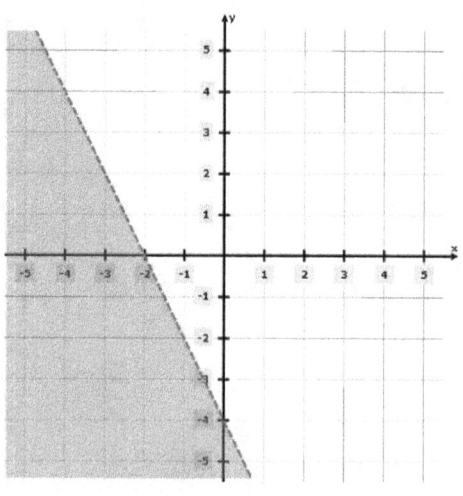

4)

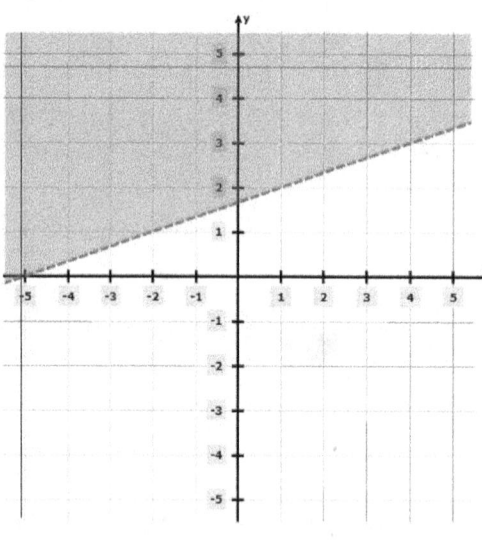

5)

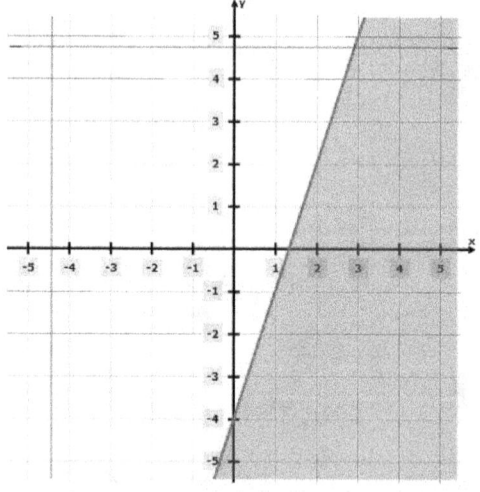

9–6 Finding Midpoint

1) (1, -7.5)
2) (2.5, -3.5)
3) (-2.5, 2)
4) (-1, 4)
5) (8, 1.5)
6) (2, 1.5)
7) (1, -4)
8) (0.5, -4.5)
9) (0, 2.5)
10) (-2.5, 2)
11) (2, -2)
12) (0.05, -0.9)
13) (4.05, 5.96)
14) (-2, - 0.5)
15) ($\frac{1}{2}$, $1\frac{1}{2}$)
16) (-4, 1)
17) (1.25, 2.5)
18) (2.849, 1.02)

9–7 Finding Distance of Two Points

1) 9
2) 7.1
3) 6.32
4) 4
5) 12.21
6) 4.12
7) 3.61
8) 12.1
9) 5.66
10) 11.04
11) 4.48
12) 17.03
13) 10.77
14) 8.25
15) 16.4
16) 10.3
17) 4.48
18) 12.1

Chapter 10: Polynomials

10–1 Classifying Polynomials

10–2 Writing Polynomials in Standard Form

10–3 Simplifying Polynomials

10–4 Adding and Subtracting Polynomials

10–5 Multiplying Monomials

10–6 Multiplying and Dividing Monomials

10–7 Multiplying a Polynomial and a Monomial

10–8 Multiplying Binomials

10–9 Factoring Trinomials

10–10 Operations with Polynomials

10-1 Classifying Polynomials

Name each polynomial by degree and number of terms.

1) x

2) $-5x^4$

3) $7x - 4$

4) -6

5) $8x + 1$

6) $9x^2 - 8x^3$

7) $2x^5$

8) $10 + 8x$

9) $5x^2 - 6x$

10) $-7x^7 + 7x^4$

11) $-8x^4 + 5x^3 - 2x^2 - 8x$

12) $4x - 9x^2 + 4x^3 - 5x^4$

13) $4x^6 + 5x^5 + x^4$

14) $-4 - 2x^2 + 8x$

15) $9x^6 - 8$

16) $7x^5 + 10x^4 - 3x + 10x^7$

17) $4x^6 - 3x^2 - 8x^4$

18) $-5x^4 + 10x - 10$

10-2 Writing Polynomials in Standard Form

Write each polynomial in standard form.

1) $3x^2 - 5x^3$

2) $3 + 4x^3 - 3$

3) $2x^2 + 1x - 6x^3$

4) $9x - 7x$

5) $12 - 7x + 9x^4$

6) $5x^2 + 13x - 2x^3$

7) $-3 + 16x - 16x$

8) $3x(x+4) - 2(x+4)$

9) $(x+5)(x-2)$

10) $3x^2 + x + 12 - 5x^2 - 2x$

11) $12x^5 + 7x^3 - 3x^5 - 8x^3$

12) $3x(2x + 5 - 2x^2)$

13) $11x(x^5 + 2x^3)$

14) $(x+6)(x+3)$

15) $(x+4)^2$

16) $(8x-7)(3x+2)$

17) $5x(3x^2 + 2x + 1)$

18) $7x(3 - x + 6x^3)$

10-3 Simplifying Polynomials

Simplify each expression.

1) $11 - 4x^2 + 3x^2 - 7x^3 + 3$

2) $2x^5 - x^3 + 8x^2 - 2x^5$

3) $(-5)(x^6 + 10) - 8(14 - x^6)$

4) $4(2x^2 + 4x^2 - 3x^3) + 6x^3 + 17$

5) $11 - 6x^2 + 5x^2 - 12x^3 + 22$

6) $2x^2 - 2x + 3x^3 + 12x - 22x$

7) $(3x - 8)(3x - 4)$

8) $(12x + 2y)^2$

9) $(12x^3 + 28x^2 + 10x + 4) \div (x + 2)$

10) $(2x + 12x^2 - 2) \div (x + 2)$

11) $(2x^3 - 1) + (3x^3 - 2x^3)$

12) $(x - 5)(x - 3)$

13) $(3x + 8)(3x - 8)$

14) $(8x^2 - 3x) - (5x - 5 - 8x^2)$

10-4 Adding and Subtracting Polynomials

Simplify each expression.

1) $(2x^3 - 2) + (2x^3 + 2)$

2) $(4x^3 + 5) - (7 - 2x^3)$

3) $(4x^2 + 2x^3) - (2x^3 + 5)$

4) $(4x^2 - x) + (3x - 5x^2)$

5) $(7x + 9) - (3x + 9)$

6) $(4x^4 - 2x) - (6x - 2x^4)$

7) $(12x - 4x^3) - (8x^3 + 6x)$

8) $(2x^3 - 8x^2) - (5x^2 - 3x^3)$

9) $(2x^2 - 6) + (9x^2 - 4x^3)$

10) $(4x^3 + 3x^4) - (x^4 - 5x^3)$

11) $(-12x^4 + 10x^5 + 2x^3) + (14x^3 + 23x^5 + 8x^4)$

12) $(13x^2 - 6x^5 - 2x) - (-10x^2 - 11x^5 + 9x)$

13) $(35 + 9x^5 - 3x^2) + (8x^4 + 3x^5) - (27 - 5x^4)$

14) $(3x^5 - 2x^3 - 4x) + (4x + 10x^4 - 23) + (x^2 - x^3 + 12)$

10-5 Multiplying Monomials

Simplify each expression.

1) $2xy^2z \times 4z^2$

2) $4xy \times x^2y$

3) $4pq^3 \times (-2p^4q)$

4) $8s^4t^2 \times st^5$

5) $12p^3 \times (-3p^4)$

6) $-4p^2q^3r \times 6pq^2r^3$

7) $(-8a^4) \times (-12a^6b)$

8) $3u^4v^2 \times (-7u^2v^3)$

9) $4u^3 \times (-2u)$

10) $-6xy^2 \times 3x^2y$

11) $12y^2z^3 \times (-y^2z)$

12) $5a^2bc^2 \times 2abc^2$

www.EffortlessMath.com

10-6 Multiplying and Dividing Monomials

Simplify.

1) $(-3x^2)(8x^4y^{12})$

2) $(7x^4y^6)(4x^3y^4)$

3) $(15x^4)(3x^9)$

4) $(12x^2y^9)(7x^9y^{12})$

5) $\dfrac{36\ x^5y^7}{4\ x^4y^5}$

6) $\dfrac{80\ x^{12}y^9}{10\ x^6y^7}$

7) $\dfrac{95\ x^{18}y^7}{5\ x^9y^2}$

8) $\dfrac{200\ x^3y^8}{40\ x^3y^7}$

9) $\dfrac{-15\ x^{17}y^{13}}{3\ x^6y^9}$

10) $\dfrac{-64\ x^8y^{10}}{8\ x^3y^7}$

10-7 Multiplying a Polynomial and a Monomial

Find each product.

1) $5(3x - 6y)$

2) $9x(2x + 4y)$

3) $8x(7x - 4)$

4) $12x(3x + 9)$

5) $11x(2x - 11y)$

6) $2x(6x - 6y)$

7) $3x(2x^2 - 3x + 8)$

8) $13x(4x + 8y)$

9) $20(2x^2 - 8x - 5)$

10) $3x(3x - 2)$

11) $6x^3(3x^2 - 2x + 2)$

12) $8x^2(3x^2 - 5xy + 7y^2)$

13) $2x^2(3x^2 - 5x + 12)$

14) $2x^3(2x^2 + 5x - 4)$

15) $5x(6x^2 - 5xy + 2y^2)$

16) $9(x^2 + xy - 8y^2)$

10-8 Multiplying Binomials

Simplify each expression.

1) $(3x - 2)(4x + 2)$

2) $(2x - 5)(x + 7)$

3) $(x + 2)(x + 8)$

4) $(x^2 + 2)(x^2 - 2)$

5) $(x - 2)(x + 4)$

6) $(x - 8)(2x + 8)$

7) $(5x - 4)(3x + 3)$

8) $(x - 7)(x - 6)$

9) $(6x + 9)(4x + 9)$

10) $(2x - 6)(5x + 6)$

11) $(x - 7)(x + 7)$

12) $(x + 4)(4x - 8)$

13) $(6x - 4)(6x + 4)$

14) $(x - 7)(x + 2)$

15) $(x - 8)(x + 8)$

16) $(3x + 3)(3x - 4)$

17) $(x + 3)(x + 3)$

18) $(x + 4)(x + 6)$

10-9 Factoring Trinomials

Factor each trinomial.

1) $x^2 - 7x + 12$

2) $x^2 + 5x - 14$

3) $x^2 - 11x - 42$

4) $6x^2 + x - 12$

5) $x^2 - 17x + 30$

6) $x^2 + 8x + 15$

7) $3x^2 + 11x - 4$

8) $x^2 - 6x - 27$

9) $10x^2 + 33x - 7$

10) $x^2 + 24x + 144$

11) $49x^2 + 28xy + 4y^2$

12) $16x^2 - 40x + 25$

13) $x^2 - 10x + 25$

14) $25x^2 - 20x + 4$

15) $x^3 + 6x^2y^2 + 9xy^3$

16) $9x^2 + 24x + 16$

17) $x^2 - 8x + 16$

18) $x^2 + 121 + 22x$

10-10 Operations with Polynomials

Find each product.

1) $3x^2 (6x - 5)$

2) $5x^2 (7x - 2)$

3) $-3 (8x - 3)$

4) $6x^3 (-3x + 4)$

5) $9 (6x + 2)$

6) $8 (3x + 7)$

7) $5 (6x - 1)$

8) $-7x^4 (2x - 4)$

9) $8 (x^2 + 2x - 3)$

10) $4 (4x^2 - 2x + 1)$

11) $2 (3x^2 + 2x - 2)$

12) $8x (5x^2 + 3x + 8)$

13) $(9x + 1)(3x - 1)$

14) $(4x + 5)(6x - 5)$

15) $(7x + 3)(5x - 6)$

16) $(3x - 4)(3x + 8)$

Name each polynomial by degree and number of terms.

17) 7

18) $2x^2 - 9x - 5$

19) $-3x$

20) $-7 + 7x^3 - x^2$

21) $-8x^4 - 8x^2$

22) $3x^6 + 3x^5 - 5x^4 - 3x^2 + 2$

Answers of Worksheets – Chapter 10

10–1 Classifying Polynomials

1) Linear monomial
2) Quartic monomial
3) Linear binomial
4) Constant monomial
5) Linear binomial
6) Quantic binomial
7) Quantic monomial
8) Linear binomial
9) Quadratic binomial
10) Seventh degree binomial
11) Quartic polynomial with four terms
12) Quartic polynomial with four terms
13) Sixth degree trinomial
14) Quadratic trinomial
15) Sixth degree binomial
16) Seventh degree polynomial with four terms
17) Sixth degree trinomial
18) Quartic trinomial

10–2 Writing Polynomials in Standard Form

1) $-5x^3 + 3x^2$
2) $4x^3$
3) $-6x^3 + 2x^2 + x$
4) $2x$
5) $9x^4 - 7x + 12$
6) $-2x^3 + 5x^2 + 13x$
7) -3
8) $3x^2 + 10x - 8$
9) $x^2 + 3x - 10$
10) $-2x^2 - x + 12$
11) $9x^5 - x^3$
12) $-6x^3 + 6x^2 + 15x$
13) $11x^6 + 22x^4$
14) $x^2 + 9x + 18$
15) $x^2 + 8x + 16$
16) $24x^2 - 5x - 14$

Arithmetic and Pre-Algebra Workbook

17) $15x^3 + 10x^2 + 5x$

18) $7x^4 - 7x^2 + 21x$

10–3 Simplifying Polynomials

1) $-7x^3 - x^2 + 14$
2) $-3x^3 + 8x^2$
3) $3x^6 - 162$
4) $-6x^3 + 24x^2 + 17$
5) $-12x^3 - x^2 + 33$
6) $3x^3 + 2x^2 - 12x$
7) $9x^2 - 36x + 32$
8) $144x^2 + 48xy + 4y^2$

9) $12x^2 + 4x + 2$
10) $12x - 22 + \frac{42}{x+2}$
11) $3x^3 - 1$
12) $x^2 - 8x + 15$
13) $9x^2 - 64$
14) $16x^2 - 8x + 5$

10–4 Adding and Subtracting Polynomials

1) $4x^3$
2) $6x^3 - 2$
3) $4x^2 - 5$
4) $-x^2 + 2x$
5) $4x$
6) $6x^4 - 8x$
7) $-12x^3 + 6x$

8) $5x^3 - 13x^2$
9) $-4x^3 + 11x^2 - 6$
10) $2x^4 + 9x^3$
11) $33x^5 - 4x^4 + 16x^3$
12) $5x^5 + 23x^2 - 11x$
13) $12x^5 + 13x^4 - 3x^2 + 8$
14) $3x^5 + 10x^4 - 3x^3 + x^2 - 11$

10–5 Multiplying Monomials

1) $8xy^2z^3$
2) $4x^3y^2$
3) $-8p^5q^4$
4) $8s^5t^7$

5) $-36p^7$
6) $-24p^3q^5r^4$
7) $96a^{10}b$
8) $-21u^6v^5$

9) $-8u^4$
10) $-18x^3y^3$
11) $-12y^4z^4$
12) $10a^3b^2c^4$

Arithmetic and Pre-Algebra Workbook

10–6 Multiplying and Dividing Monomials

1) $-24x^6y^{12}$
2) $28x^7y^{10}$
3) $45x^{13}$
4) $84x^{11}y^{21}$
5) $9xy^2$
6) $8x^6y^2$
7) $19x^9y^5$
8) $5y$
9) $-5x^{11}y^4$
10) $-8x^5y^3$

10–7 Multiplying a Polynomial and a Monomial

1) $15x - 30y$
2) $18x^2 + 36xy$
3) $56x^2 - 32x$
4) $36x^2 + 108x$
5) $22x^2 - 121xy$
6) $12x^2 - 12xy$
7) $6x^3 - 9x^2 + 24x$
8) $52x^2 + 104xy$
9) $40x^2 - 160x - 100$
10) $9x^2 - 6x$
11) $18x^5 - 12x^4 + 12x^3$
12) $24x^4 - 40x^3y + 56y^2x^2$
13) $6x^4 - 10x^3 + 24x^2$
14) $4x^5 + 10x^4 - 8x^3$
15) $30x^3 - 25x^2y + 10xy^2$
16) $9x^2 + 9xy - 72y^2$

10–8 Multiplying Binomials

1) $12x^2 - 2x - 4$
2) $2x^2 + 9x - 35$
3) $x^2 + 10x + 16$
4) $x^4 - 4$
5) $x^2 + 2x - 8$
6) $2x^2 - 8x - 64$
7) $15x^2 + 3x - 12$
8) $x^2 - 13x + 42$
9) $24x^2 + 90x + 81$
10) $10x^2 - 18x - 36$
11) $x^2 - 49$
12) $4x^2 + 8x - 32$
13) $36x^2 - 16$
14) $x^2 - 5x - 14$
15) $x^2 - 64$
16) $9x^2 - 3x - 12$
17) $x^2 + 6x + 9$
18) $x^2 + 10x + 24$

10–9 Factoring Trinomials

1) $(x - 3)(x - 4)$
2) $(x - 2)(x + 7)$
3) $(x + 3)(x - 14)$
4) $(2x + 3)(3x - 4)$
5) $(x - 15)(x - 2)$
6) $(x + 3)(x + 5)$

www.EffortlessMath.com

7) $(3x + 1)(x - 4)$

8) $(x - 9)(x + 3)$

9) $(5x - 1)(2x + 7)$

10) $(x + 12)(x + 12)$

11) $(7x + 2y)(7x + 2y)$

12) $(4x - 5)(4x - 5)$

13) $(x - 5)(x - 5)$

14) $(5x - 2)(5x - 2)$

15) $x(x^2 + 6xy^2 + 9y^3)$

16) $(3x + 4)(3x + 4)$

17) $(x - 4)(x - 4)$

18) $(x + 11)(x + 11)$

10–10 Operations with Polynomials

1) $18x^3 - 15x^2$

2) $35x^3 - 10x^2$

3) $-24x + 9$

4) $-18x^4 + 24x^3$

5) $54x + 18$

6) $24x + 56$

7) $30x - 5$

8) $-14x^5 + 28x^4$

9) $8x^2 + 16x - 24$

10) $16x^2 - 8x + 4$

11) $6x^2 + 4x - 4$

12) $40x^3 + 24x^2 + 64x$

13) $27x^2 - 6x - 1$

14) $24x^2 + 10x - 25$

15) $35x^2 - 27x - 18$

16) $9x^2 + 12x - 32$

17) Constant monomial

18) Quadratic

19) Linear monomial

20) Cubic trinomial

21) Quartic binomial

22) Sixth degree polynomial with five terms

Chapter 11: Exponents and Radicals

11–1 Multiplication Property of Exponents

11–2 Division Property of Exponents

11–3 Powers of Products and Quotients

11–4 Zero and Negative Exponents

11–5 Negative Exponents and Negative Bases

11–6 Writing Scientific Notation

11–7 Square Roots

11-1 Multiplication Property of Exponents

Simplify.

1) $4^2 \cdot 4^2$

2) $2 \cdot 2^2 \cdot 2^2$

3) $3^2 \cdot 3^2$

4) $3x^3 \cdot x$

5) $12x^4 \cdot 3x$

6) $6x \cdot 2x^2$

7) $5x^4 \cdot 5x^4$

8) $6x^2 \cdot 6x^3y^4$

9) $7x^2y^5 \cdot 9xy^3$

10) $7xy^4 \cdot 4x^3y^3$

11) $(2x^2)^2$

12) $3x^5y^3 \cdot 8x^2y^3$

13) $7x^3 \cdot 10y^3x^5 \cdot 8yx^3$

14) $(x^4)^3$

15) $(2x^2)^4$

16) $(x^2)^3$

17) $(6x)^2$

18) $3x^4y^5 \cdot 7x^2y^3$

11-2 Division Property of Exponents

Simplify.

1) $\dfrac{5^5}{5}$

2) $\dfrac{3}{3^5}$

3) $\dfrac{2^2}{2^3}$

4) $\dfrac{2^4}{2^2}$

5) $\dfrac{x}{x^3}$

6) $\dfrac{3x^3}{9x^4}$

7) $\dfrac{2x^{-5}}{9x^{-2}}$

8) $\dfrac{21x^8}{7x^3}$

9) $\dfrac{7x^6}{4x^7}$

10) $\dfrac{6x^2}{4x^3}$

11) $\dfrac{5x}{10x^3}$

12) $\dfrac{3x^3}{2x^5}$

13) $\dfrac{12x^3}{14x^6}$

14) $\dfrac{12x^3}{9y^8}$

15) $\dfrac{25xy^4}{5x^6y^2}$

16) $\dfrac{2x^4}{7x}$

17) $\dfrac{16x^2y^8}{4x^3}$

18) $\dfrac{12x^4}{15x^7y^9}$

19) $\dfrac{12yx^4}{10yx^8}$

20) $\dfrac{16x^4y}{9x^8y^2}$

21) $\dfrac{5x^8}{20x^8}$

11-3 Powers of Products and Quotients

Simplify.

1) $(2x^3)^4$

2) $(4xy^4)^2$

3) $(5x^4)^2$

4) $(11x^5)^2$

5) $(4x^2y^4)^4$

6) $(2x^4y^4)^3$

7) $(3x^2y^2)^2$

8) $(3x^4y^3)^4$

9) $(2x^6y^8)^2$

10) $(12x\ 3x)^3$

11) $(2x^9\ x^6)^3$

12) $(5x^{10}y^3)^3$

13) $(4x^3\ x^2)^2$

14) $(3x^3\ 5x)^2$

15) $(10x^{11}y^3)^2$

16) $(9x^7\ y^5)^2$

17) $(4x^4y^6)^5$

18) $(4x^4)^2$

19) $(3x\ 4y^3)^2$

20) $(9x^2y)^3$

21) $(12x^2y^5)^2$

11-4 Zero and Negative Exponents

Evaluate the following expressions.

1) 0.8^{-2}

2) 0.2^{-4}

3) 10^{-2}

4) 0.5^{-3}

5) 22^{-1}

6) 9^{-1}

7) 3^{-2}

8) 4^{-2}

9) 5^{-2}

10) 35^{-1}

11) 6^{-3}

12) 0^{15}

13) 10^{-9}

14) 3^{-4}

15) 5^{-2}

16) $2^{\ 4}$

17) 7^{-3}

18) 8^{-1}

19) 7^{-3}

20) 2^{-4}

21) $(\frac{2}{3})^{-2}$

22) $(\frac{1}{5})^{-3}$

23) $(\frac{5}{10})^{-8}$

24) $(\frac{2}{5})^{-3}$

11-5 Negative Exponents and Negative Bases

Simplify.

1) -6^{-1}

2) $-4x^{-3}$

3) $-\dfrac{5x}{x^{-3}}$

4) $-\dfrac{a^{-3}}{b^{-2}}$

5) $-\dfrac{5}{x^{-3}}$

6) $\dfrac{7b}{-9c^{-4}}$

7) $-\dfrac{5n^{-2}}{10p^{-3}}$

8) $\dfrac{4ab^{-2}}{-3c^{-2}}$

9) $-12x^2y^{-3}$

10) $\left(-\dfrac{1}{3}\right)^{-2}$

11) $\left(-\dfrac{3}{4}\right)^{-2}$

12) $\left(\dfrac{3a}{2c}\right)^{-2}$

13) $\left(-\dfrac{5x}{3yz}\right)^{-3}$

14) $-\dfrac{2x}{a^{-4}}$

11-6 Writing Scientific Notation

Write each number in scientific notation.

1) 91×10^3

2) 60

3) 2000000

4) 0.0000006

5) 354000

6) 0.000325

7) 2.5

8) 0.00023

9) 56000000

10) 2000000

11) 78000000

12) 0.0000022

13) 0.00012

14) 0.004

15) 78

16) 1600

17) 1450

18) 130000

19) 60

20) 0.113

21) 0.02

11-7 Square Roots

Find the value each square root.

1) $\sqrt{1}$

2) $\sqrt{4}$

3) $\sqrt{9}$

4) $\sqrt{25}$

5) $\sqrt{16}$

6) $\sqrt{49}$

7) $\sqrt{36}$

8) $\sqrt{0}$

9) $\sqrt{64}$

10) $\sqrt{81}$

11) $\sqrt{121}$

12) $\sqrt{225}$

13) $\sqrt{144}$

14) $\sqrt{100}$

15) $\sqrt{256}$

16) $\sqrt{289}$

17) $\sqrt{324}$

18) $\sqrt{400}$

19) $\sqrt{900}$

20) $\sqrt{529}$

21) $\sqrt{90}$

Arithmetic and Pre-Algebra Workbook

Answers of Worksheets – Chapter 11

11–1 Multiplication Property of Exponents

1) 4^4
2) 2^5
3) 3^4
4) $3x^4$
5) $36x^5$
6) $12x^3$
7) $25x^8$
8) $36x^5y^4$
9) $63x^3y^8$
10) $28x^4y^7$
11) $4x^4$
12) $24x^7y^6$
13) $560x^{11}y^4$
14) x^{12}
15) $16x^8$
16) x^6
17) $36x^2$
18) $21x^6y^8$

11–2 Division Property of Exponents

1) 5^4
2) $\frac{1}{3^4}$
3) $\frac{1}{2}$
4) 2^2
5) $\frac{1}{x^2}$
6) $\frac{1}{3x}$
7) $\frac{2}{9x^3}$
8) $3x^5$
9) $\frac{4}{7x}$
10) $\frac{2}{3x}$
11) $\frac{1}{2x^2}$
12) $\frac{2}{3x^2}$
13) $\frac{6}{7x^3}$
14) $\frac{3x^3}{4y^8}$
15) $\frac{5y^2}{x^5}$
16) $\frac{2x^3}{7}$
17) $\frac{4y^8}{x}$
18) $\frac{4}{5x^3y^9}$
19) $\frac{6}{5x^4}$
20) $\frac{16}{9x^4y}$
21) $\frac{1}{4}$

11–3 Powers of Products and Quotients

1) $16x^{12}$
2) $16x^2y^8$
3) $25x^8$
4) $121x^{10}$
5) $256x^8y^{16}$
6) $8x^{12}y^{12}$
7) $9x^4y^4$
8) $81x^{16}y^{12}$
9) $4x^{12}y^{16}$

www.EffortlessMath.com

Arithmetic and Pre-Algebra Workbook

10) $46.656x^6$

11) $8x^{45}$

12) $125x^{30}y^9$

13) $16x^{10}$

14) $225x^8$

15) $100x^{22}y^6$

16) $81x^{14}y^{10}$

17) $1,024x^{20}y^{30}$

18) $16x^8$

19) $144x^2y^6$

20) $729x^6y^3$

21) $144x^4y^{10}$

11–4 Zero and Negative Exponents

1) $\dfrac{1}{64}$

2) $\dfrac{1}{16}$

3) $\dfrac{1}{100}$

4) $\dfrac{1}{125}$

5) $\dfrac{1}{22}$

6) $\dfrac{1}{9}$

7) $\dfrac{1}{9}$

8) $\dfrac{1}{16}$

9) $\dfrac{1}{25}$

10) $\dfrac{1}{35}$

11) $\dfrac{1}{216}$

12) 0

13) $\dfrac{1}{1000000000}$

14) $\dfrac{1}{81}$

15) $\dfrac{1}{25}$

16) $\dfrac{1}{8}$

17) $\dfrac{1}{27}$

18) $\dfrac{1}{8}$

19) $\dfrac{1}{343}$

20) $\dfrac{1}{36}$

21) $\dfrac{9}{4}$

22) 125

23) 256

24) $\dfrac{125}{8}$

11–5 Negative Exponents and Negative Bases

1) $-\dfrac{1}{6}$

2) $-\dfrac{4}{x^3}$

3) $-5x^4$

4) $-\dfrac{b^2}{a^3}$

5) $-5x^3$

6) $-\dfrac{7bc^4}{9}$

7) $-\dfrac{p^3}{2n^2}$

8) $-\dfrac{4ac^2}{3b^2}$

9) $-\dfrac{12x^2}{y^3}$

10) 9

11) $\dfrac{16}{9}$

12) $\dfrac{4c^2}{9a^2}$

13) $-\dfrac{27y^3z^3}{125x^3}$

14) $-2xa^4$

Arithmetic and Pre-Algebra Workbook

11–6 Writing Scientific Notation

1) 9.1×10^4
2) 6×10^1
3) 2×10^6
4) 6×10^{-7}
5) 3.54×10^5
6) 3.25×10^{-4}
7) 2.5×10^0
8) 2.3×10^{-4}
9) 5.6×10^7
10) 2×10^6
11) 7.8×10^7
12) 2.2×10^{-6}
13) 1.2×10^{-4}
14) 4×10^{-3}
15) 7.8×10^1
16) 1.6×10^3
17) 1.45×10^3
18) 1.3×10^5
19) 6×10^1
20) 1.13×10^{-1}
21) 2×10^{-2}

11–7 Square Roots

1) 1
2) 2
3) 3
4) 5
5) 4
6) 7
7) 6
8) 0
9) 8
10) 9
11) 11
12) 15
13) 12
14) 10
15) 16
16) 17
17) 18
18) 20
19) 30
20) 23
21) $3\sqrt{10}$

Chapter 12: Plane Figures

12–1 Transformations: Translations, Rotations, and Reflections

12–2 The Pythagorean Theorem

12–3 Classifying Triangles and Quadrilaterals

12–4 Area of Triangles

12–5 Perimeter of Polygons

12–6 Area and Circumference of Circles

12–7 Area of Squares, Rectangles, and Parallelograms

12–8 Area of Trapezoids

12-1 Transformations: Translations, Rotations, and Reflections

Graph.

1) translation: 4 units right and 1 unit down

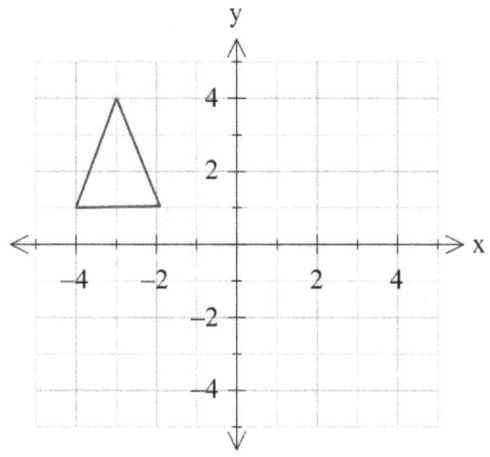

2) translation: 4 units right and 2 unit up

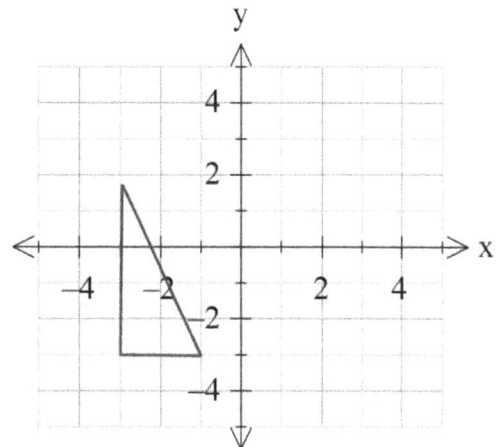

3) rotation 90° counterclockwise about the origin

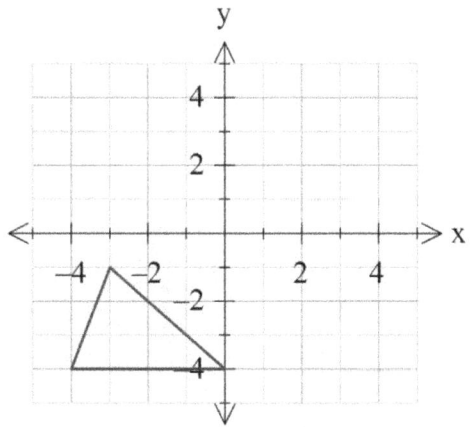

4) rotation 180° about the origin

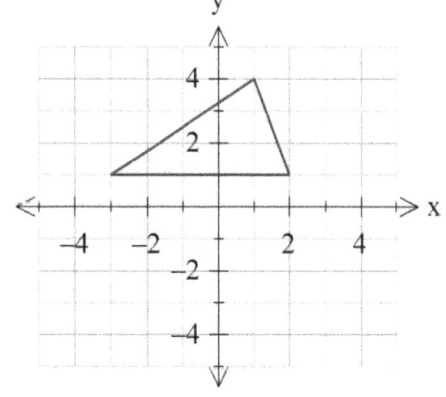

12-2 The Pythagorean Theorem

Do the following lengths form a right triangle?

1)

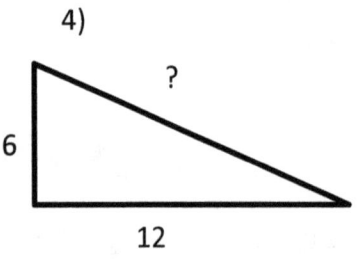

Sides: 8, 10, 6

2)

Sides: 3, 4, 5

3)

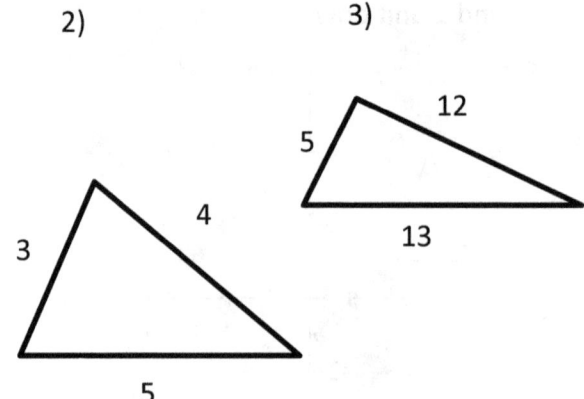

Sides: 5, 12, 13

Find each missing length to the nearest tenth.

4)

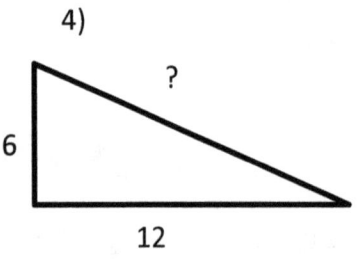

Sides: 6, 12, ?

5)

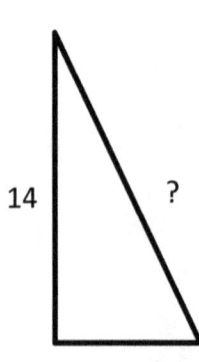

Sides: 14, 8, ?

6)

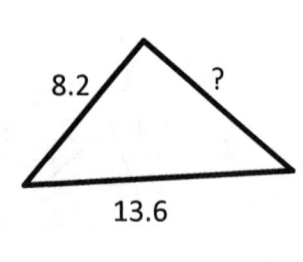

Sides: 8.2, 13.6, ?

12-3 Classifying Triangles and Quadrilaterals

Classify each triangle by its angles and sides.

1)

2)

3)

4)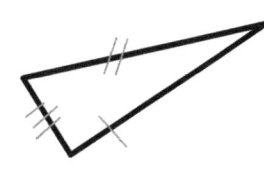

Classify each quadrilateral with the name that best describes it.

5)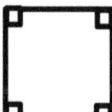

7)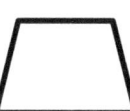

6)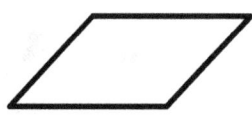

8)

www.EffortlessMath.com

12-4 Area of Triangles

Find the area of each.

1)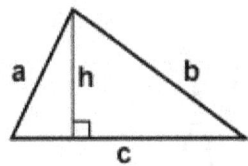

c = 9 mi

h = 3.7 mi

2)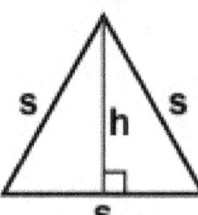

s = 14 m

h = 8 m

3)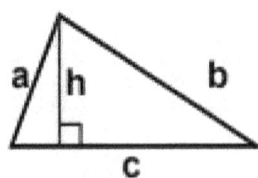

a = 5 m

b = 11 m

c = 14 m

h = 4 m

4)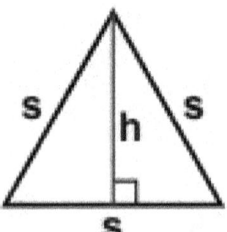

s = 16 m

h = 12.1 m

12-5 Perimeter of Polygons

Find the perimeter of each shape.

1)
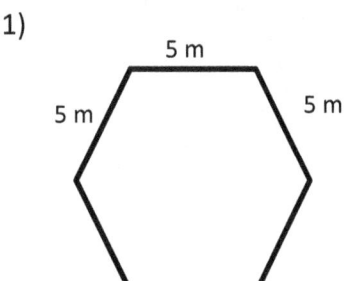

2)
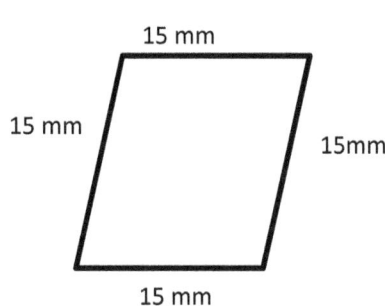

3)
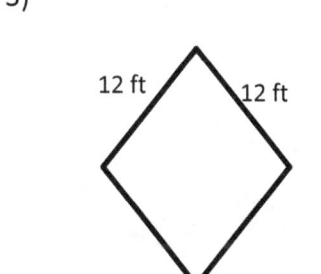

4)
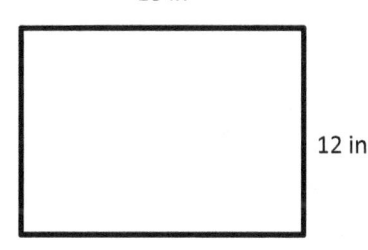

5)
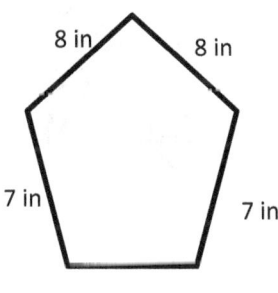

6)
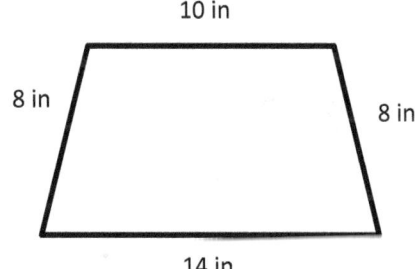

www.EffortlessMath.com

12-6 Area and Circumference of Circles

Find the area and circumference of each.

1)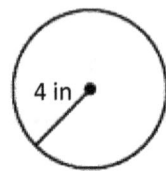

2)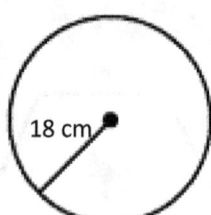

3)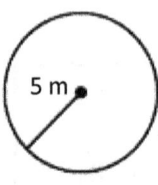

4)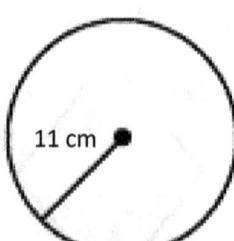

5)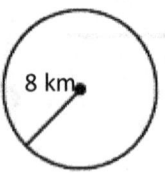

6)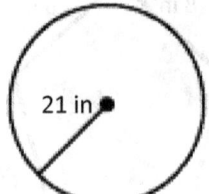

12-7 Area of Squares, Rectangles, and Parallelograms

Find the area of each.

1)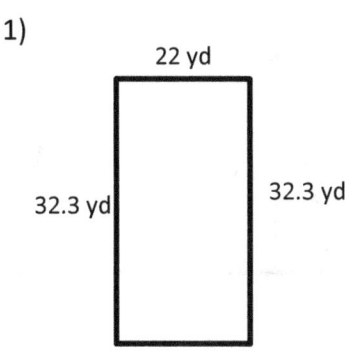

2)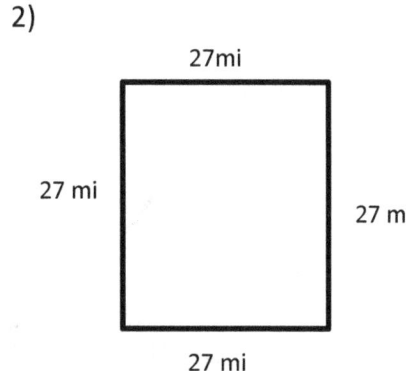

3)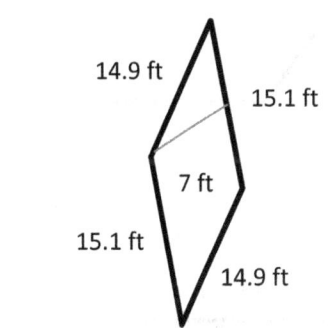

4)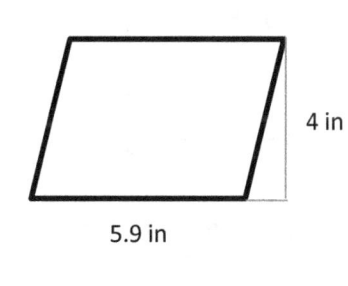

5)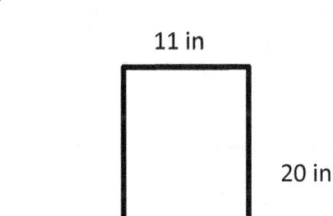

6)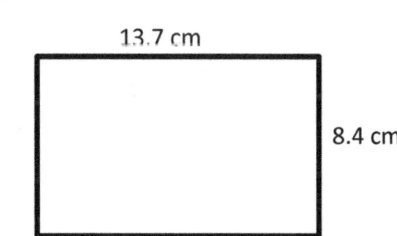

12-8 Area of Trapezoids

Calculate the area for each trapezoid.

1)
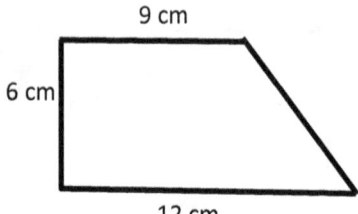

2)
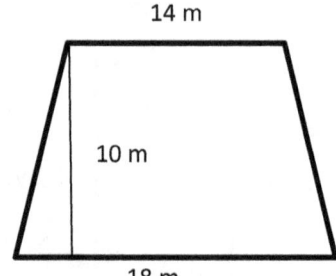

3)
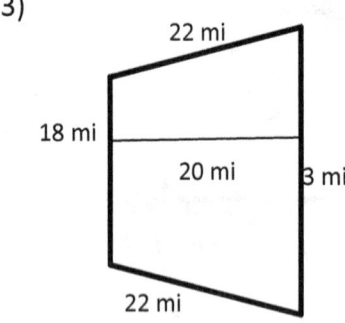

4)
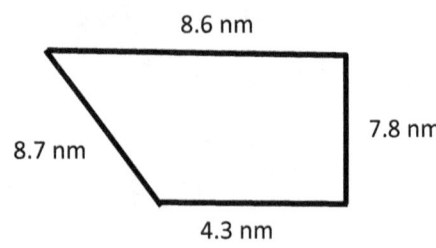

5)
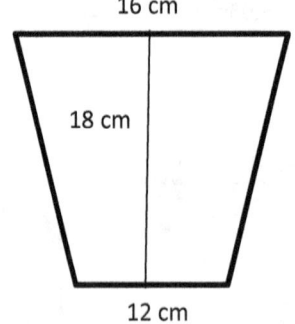

6)
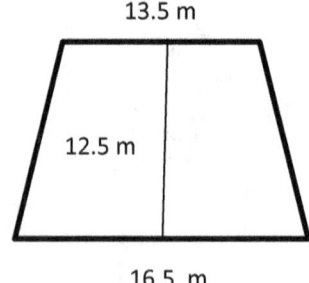

Arithmetic and Pre-Algebra Workbook

Answers of Worksheets – Chapter 12

12–1 Transformations: Translations, Rotations, and Reflections

1)

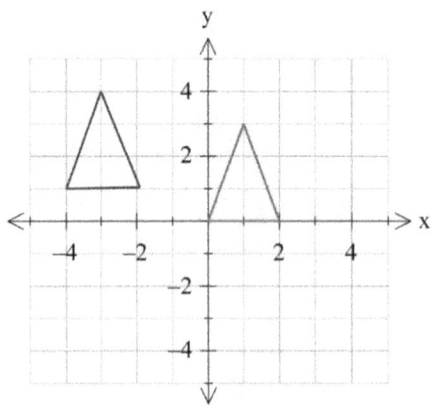

2)

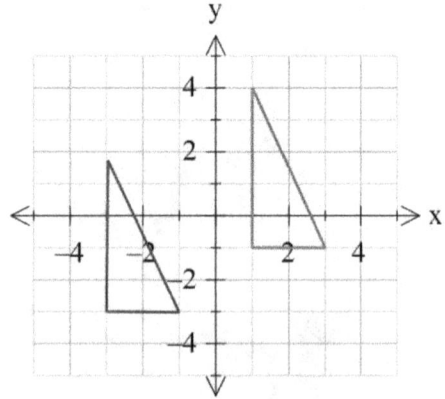

3) rotation 90° counterclockwise about the origin

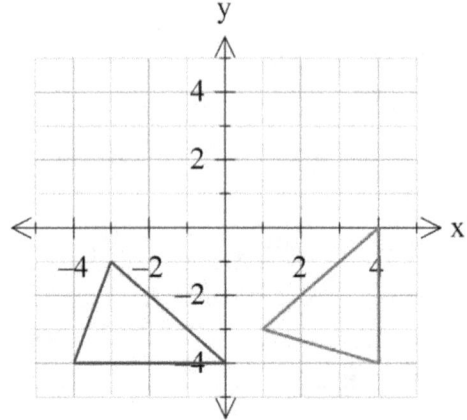

4) rotation 180° about the origin

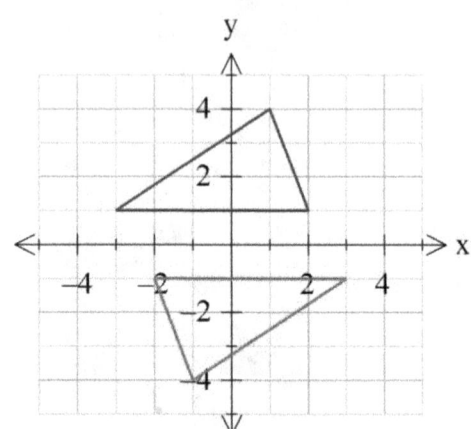

Arithmetic and Pre-Algebra Workbook

12–2 The Pythagorean Theorem

1) yes
2) yes
3) yes
4) 17
5) 26
6) 13

12–3 Classifying Triangles and Quadrilaterals

1) Right isosceles
2) Equilateral
3) Acute isosceles
4) Obtuse scalene
5) Square
6) Trapezoid
7) Rhombus
8) Trapezoid

12–4 Area of Triangles

1) 16.65 mi^2
2) 24.49 m^2
3) 84.8 m^2
4) 45.5 yd^2
5) 136.4 km^2
6) 110.85 m^2

12–5 Perimeter of Polygons

1) 30 m
2) 60 mm
3) 48 ft
4) 60 in
5) 35 in
6) 40 in

12–6 Area and Circumference of Circles

1) Area: 50.27 in^2, Circumference: 25.12 in
2) Area: 1,017.36 cm^2, Circumference: 113.04 cm
3) Area: 78.5 m^2, Circumference: 31.4 m
4) Area: 379.94 cm^2, Circumference: 69.08 cm
5) Area: 200.96 km^2, Circumference: 50.2 km
6) Area: 1,384.74 km^2, Circumference: 131.88 km

Arithmetic and Pre-Algebra Workbook

12–7 Area of Squares, Rectangles, and Parallelograms

1) 710.6 yd^2
2) 729 mi^2
3) 105.7 ft^2
4) 23.6 in^2
5) 220 in^2
6) 115.08 cm^2

12–8 Area of Trapezoids

1) 63 cm^2
2) 192 m^2
3) 451 mi^2
4) 50.31 nm^2
5) 280 cm^2
6) 180 m^2

Chapter 13: Solid Figures

13–1 Classifying Solids

13–2 Volume of Cubes and Rectangle Prisms

13–3 Surface Area of Cubes

13–4 Surface Area of a Prism

13–5 Surface Area of a Cylinder

13–6 Surface Area of Pyramids and Cones

13–7 Surface Area of a Sphere

13–8 Volume of a Pyramid and Cone

13–9 Volume of a Sphere

13-1 Classifying Solids

Identify the names of the following shapes.

1)
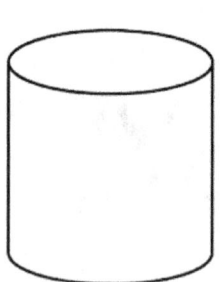

2)
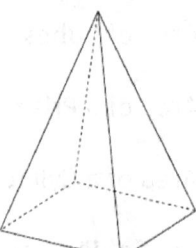

3)

4)
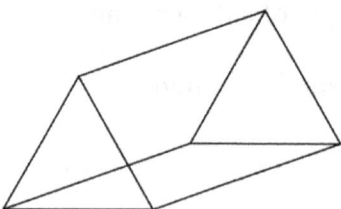

5)
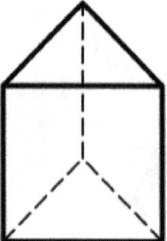

6)

Arithmetic and Pre-Algebra Workbook

13-2 Volume of Cubes and Rectangle Prisms

Find the volume of each of the rectangular prisms.

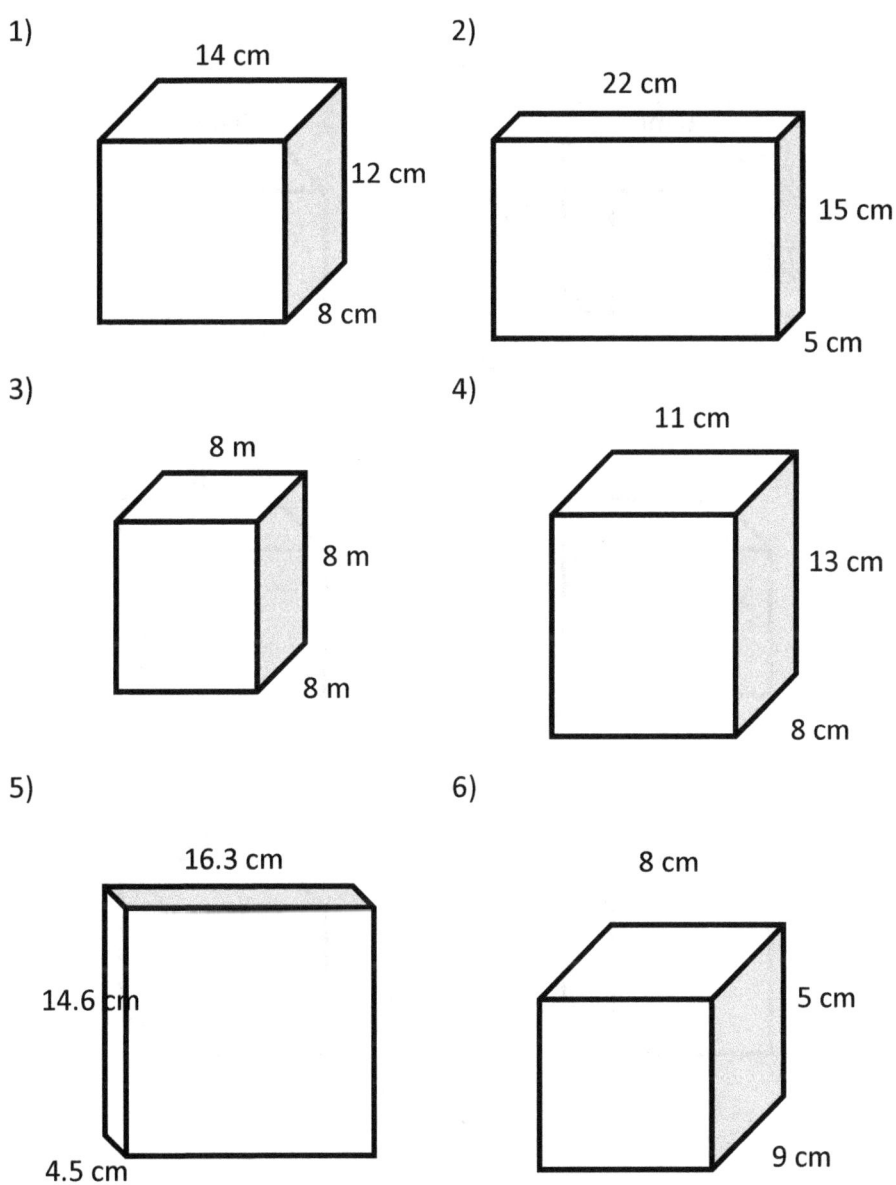

www.EffortlessMath.com

13-3 Surface Area of Cubes

Find the surface of each prism.

1)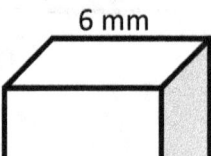

2)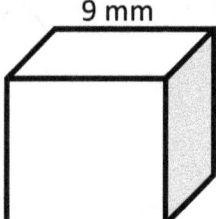

3)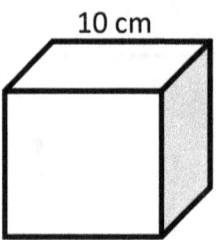

4)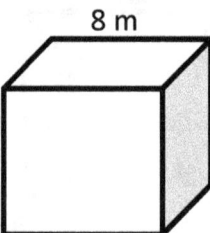

5)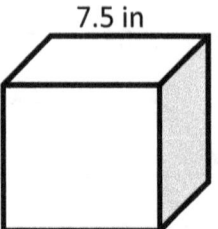

6)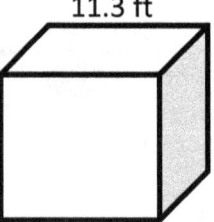

Arithmetic and Pre-Algebra Workbook

13-4 Surface Area of a Prism

Find the surface of each prism.

1)

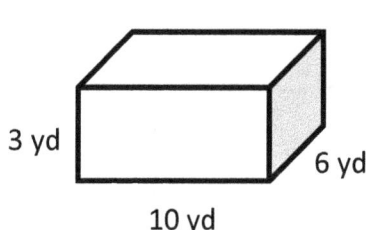

2)

7 mm
7 mm
7 mm

3)

8 in
13.2 in
6.7 in

4)

17 cm
17 cm
11 cm

5)

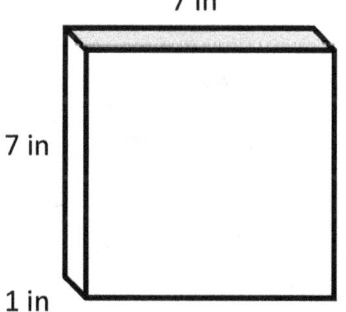

6)

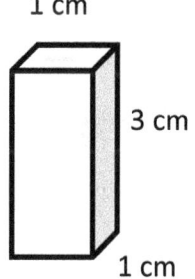

www.EffortlessMath.com

Arithmetic and Pre-Algebra Workbook

13-5 Surface Area of a Cylinder

Find the surface of each cylinder.

1)

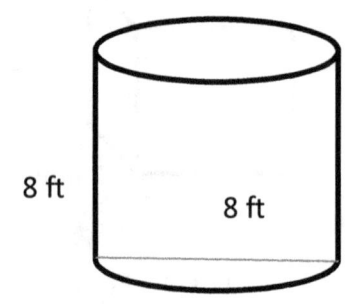

2)

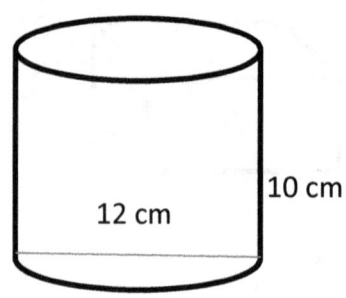

3)

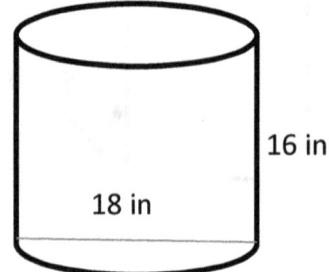

4)

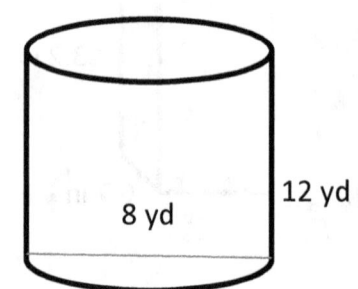

5)

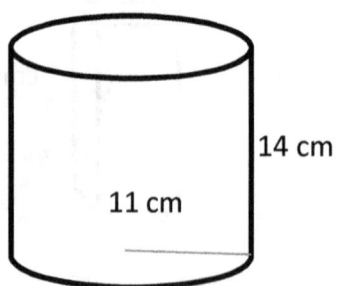

6)

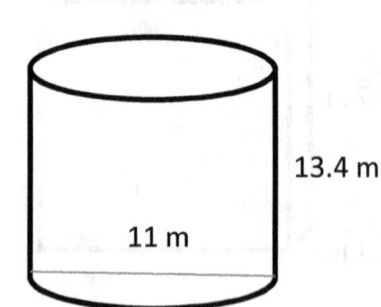

13-6 Surface Area of Pyramids and Cones

Find the surface area of each figure.

1)

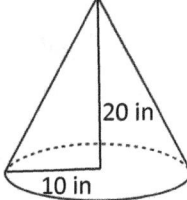

2)

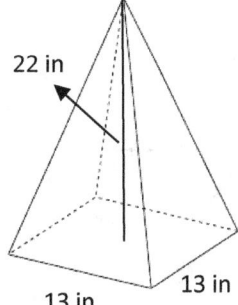

3)

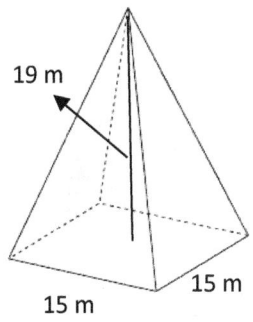

4)

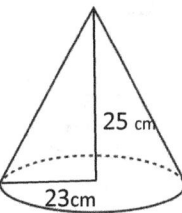

5)

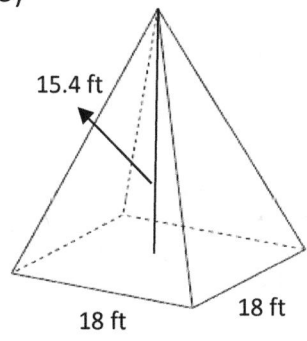

6)

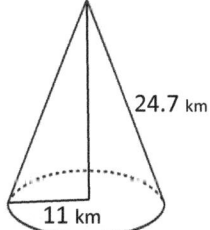

13-7 Surface Area of a Sphere

Find the surface area of each figure. Round your answer to near tenth.

1)

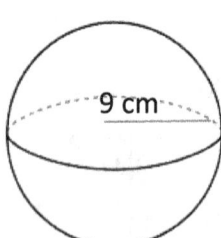

2)

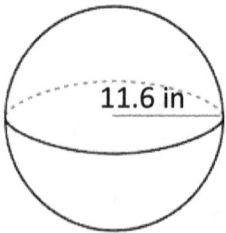

3)

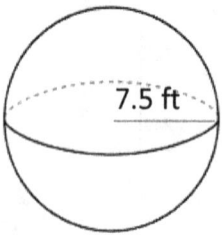

4)

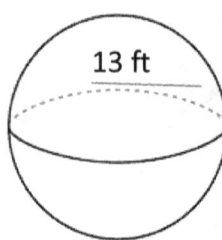

5)

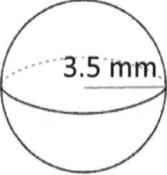

6)

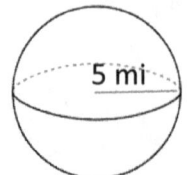

13-8 Volume of a Pyramid and Cone

Find the volume of each figure.

1)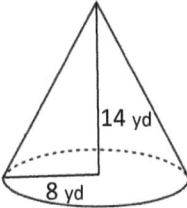

2)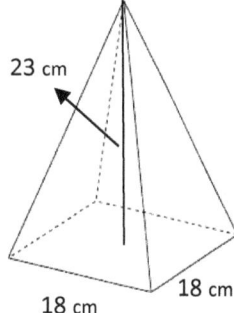

3)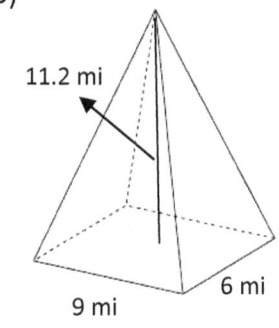

4)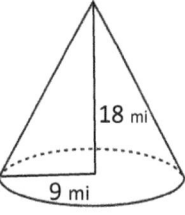

5)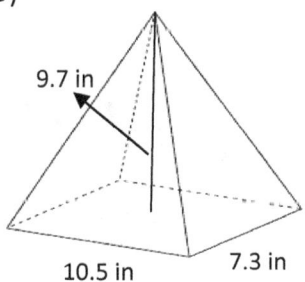

6)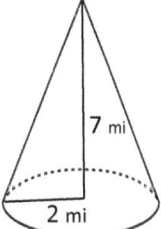

13-9 Volume of a Sphere

Find the volume of each figure.

1)

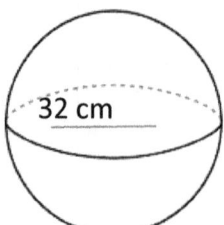
32 cm

2)

25 cm

3)

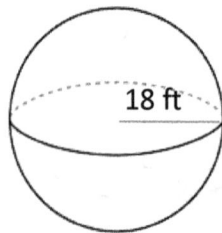
18 ft

4)

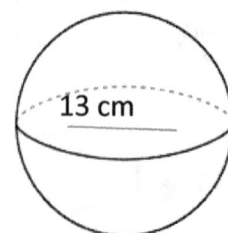
13 cm

5)

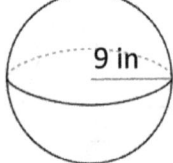
9 in

6)

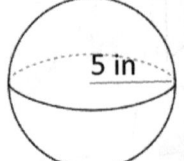
5 in

Arithmetic and Pre-Algebra Workbook

Answers of Worksheets – Chapter 13

13–1 Classifying Solids

1) Cylinder
2) Triangular pyramid
3) Rectangular prism
4) Triangular pyramid
5) Triangular prism
6) Cylinder

13–2 Volume of Cubes and Rectangle Prisms

1) 1344 cm^3
2) 1650 cm^3
3) 512 m^3
4) 1144 cm^3
5) 1070.91 m^3
6) 360 m^3

13–3 Surface Area of a Cube

1) 216
2) 486
3) 600
4) 384
5) 337.5
6) 766.14

13–4 Surface Area of a Prism

1) 216 yd^2
2) 294 mm^2
3) 495.28 in^2
4) 1326 cm^2
5) 126 in^2
6) 14 cm^2

13–5 Surface Area of a Cylinder

1) 301.714 ft^2
2) 603.43 cm^2
3) 1414.29 in^2
4) 402.12 yd^2
5) 1728.57 cm^2
6) 653.4 m^2

13–6 Surface Area of Pyramids and Cones

1) 942.48 in^3
2) 596.444 in^2
3) 612.8 m^3
4) 468.32 m^3
5) 642.1334 ft^2
6) 233.7 km^2

13–7 Surface Area of a Sphere

1) 1018.29 cm²
2) 1691.61 in²
3) 707.1429 ft²
4) 2124.57 ft²
5) 154 mm²
6) 314.29 mi²

13–8 Volume of a Pyramid and Cone

1) 938.3 yd³
2) 2484 cm³
3) 201.6 mi³
4) 1526.8 mi³
5) 247.835 in³
6) 29.3 mi³

13–9 Volume of a Sphere

1) 137,258.28 cm³
2) 65,449.85 cm³
3) 24,429.024 ft³
4) 9,202.78 cm³
5) 3,053.63 in³
6) 523.6 in³

Chapter 14: Statistics

14–1 Mean, Median, Mode, and Range of the Given Data

14–2 First Quartile, Second Quartile and Third Quartile of the Given Data

14–3 Bar Graph

14–4 Box and Whisker Plots

14–5 Stem–And–Leaf Plot

14–6 The Pie Graph or Circle Graph

14–7 Scatter Plots

Arithmetic and Pre-Algebra Workbook

14-1 Mean, Median, Mode, and Range of the Given Data

Write Mean, Median, Mode, and Range of the Given Data.

1) 7, 2, 5, 1, 1, 2

2) 2, 2, 2, 3, 6, 3, 7, 4

3) 9, 4, 3, 1, 7, 9, 4, 6, 4

4) 8, 4, 2, 4, 3, 2, 4, 5

5) 8, 5, 7, 5, 7, 9, 8

6) 5, 1, 4, 4, 9, 2, 9, 2, 5, 1

7) 4, 1, 5, 9, 7, 7, 5, 4, 3, 5

8) 7, 5, 4, 9, 6, 7, 7, 5, 2

9) 2, 5, 5, 6, 2, 4, 7, 6, 4, 9

10) 10, 5, 2, 5, 4, 5, 8, 10

11) 5, 1, 5, 2, 2

12) 2, 3, 5, 9, 6

14-2 First Quartile, Second Quartile and Third Quartile of the Given Data

Find First Quartile, Second Quartile and Third Quartile of the Given Data.

1) 65, 8, 35, 54, 29, 42, 14, 73, 11

2) 14, 64, 30, 20, 72, 57

3) 99, 37, 83, 62, 74, 49, 59, 40

4) 33, 14, 47, 29, 52, 63, 20, 39, 74, 48

5) 23, 10, 13, 30, 26, 8, 25, 18

6) 35, 60, 20, 80, 95, 15, 40, 85, 75

14-3 Bar Graph

Graph the given information as a bar graph.

Day	Hot dogs sold
Monday	90
Tuesday	70
Wednesday	30
Thursday	20
Friday	60

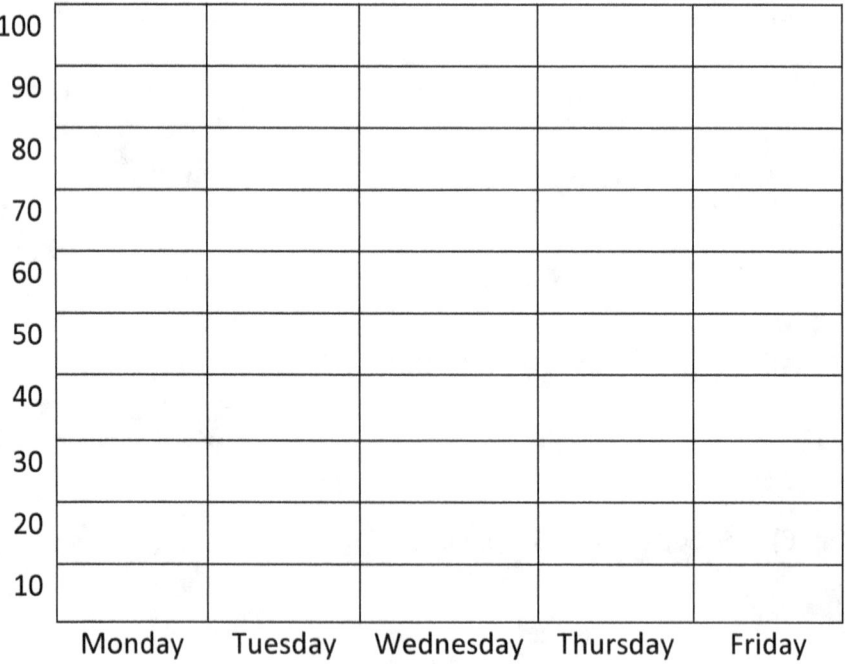

Arithmetic and Pre-Algebra Workbook

14-4 Box and Whisker Plots

Make box and whisker plots for the given data.

1) 73, 84, 86, 95, 68, 67, 100, 94, 77, 80, 62, 79

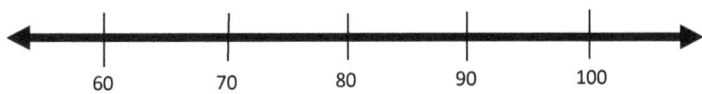

2) 11, 17, 22, 18, 23, 2, 3, 16, 21, 7, 8, 15, 5

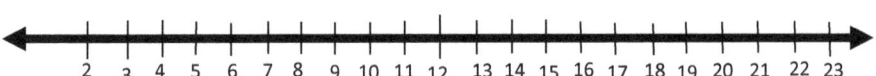

3) 20, 12, 1, 24, 14, 23, 8, 2, 22, 12, 3

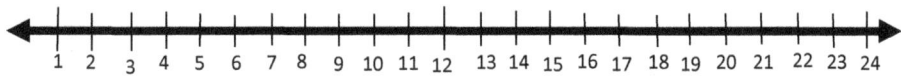

www.EffortlessMath.com

14-5 Stem-And-Leaf Plot

Make stem ad leaf plots for the given data.

1) 74, 88, 97, 72, 79, 86, 95, 79, 83, 91

   Key: 8 / 6 =

   Stem | Leaf plot

2) 37, 48, 26, 33, 49, 26, 19, 26, 48

   Key: 3 / 7 =

   Stem | Leaf plot

3) 58, 41, 42, 67, 54, 65, 65, 54, 69, 53

   Key: 6 / 5 =

   Stem | Leaf plot

Arithmetic and Pre-Algebra Workbook

14-6 The Pie Graph or Circle Graph

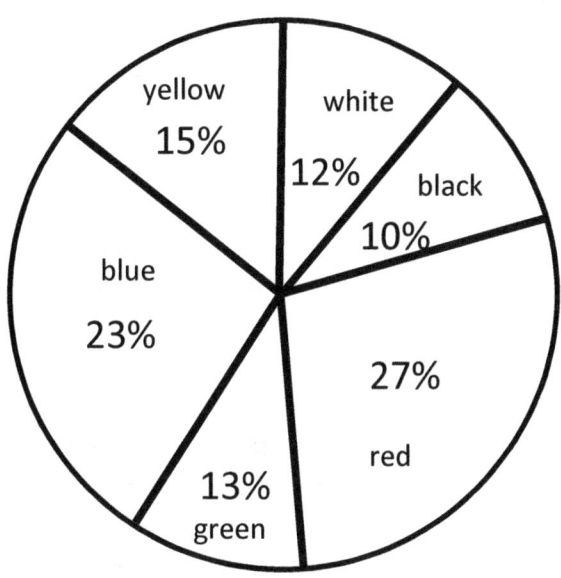

Favorite colors

1) Which color is most?

2) What percentage of pie graph is yellow?

3) Which color is least?

4) What percentage of pie graph is blue?

5) What percentage of pie graph is green?

14-7 Scatter Plots

Construct a scatter plot.

X	Y
1	20
2	40
3	50
4	60

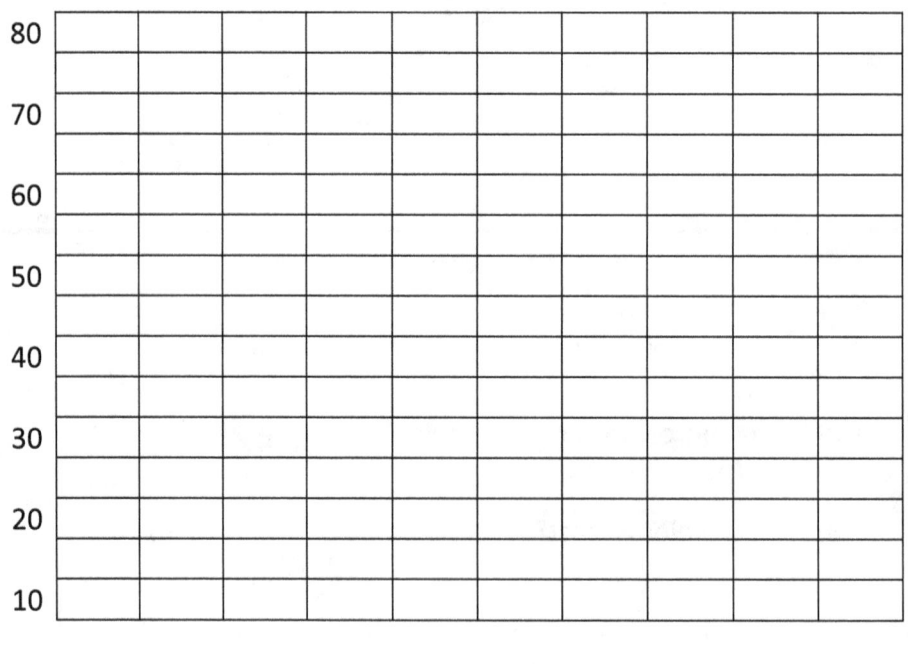

Answers of Worksheets – Chapter 14

14–1 Mean, Median, Mode, and Range of the Given Data

1) mean: 3, median: 2, mode: 2, range: 6
2) mean: 3.625, median: 3, mode: 2, range: 5
3) mean: 5.22, median: 4, mode: 4, range: 8
4) mean: 4, median: 4, mode: 4, range: 6
5) mean: 7, median: 7, mode: 5, 7, 8, range: 4
6) mean: 4.2, median: 4, mode: 1,2,4,5,9, range: 8
7) mean: 5, median: 5, mode: 5, range: 8
8) mean: 5.78, median: 6, mode: 7, range: 7
9) mean: 5, median: 5, mode: 2, 4, 5, 6, range: 7
10) mean: 6.125, median: 5, mode: 5, range: 8
11) mean: 3, median: 2, mode: 2, 5, range: 4
12) mean: 5, median: 5, mode: none, range: 7

14–2 First Quartile, Second Quartile and Third Quartile of the Given Data

1) First quartile: 12.5, second quartile: 35, third quartile: 59.5
2) First quartile: 20, second quartile: 43.5, third quartile: 64
3) First quartile: 44.5, second quartile: 60.5, third quartile: 78.5
4) First quartile: 29, second quartile: 43, third quartile: 52
5) First quartile: 11.5, second quartile: 20.5, third quartile: 25.5
6) First quartile: 27.5, second quartile: 60, third quartile: 82.5

14–3 Box and Whisker Plots

1) 73, 84, 86, 95, 68, 67, 100, 94, 77, 80, 62, 79

Maximum: 100, Minimum: 62, Q_1: 70.5, Q_2: 79.5, Q_3: 90

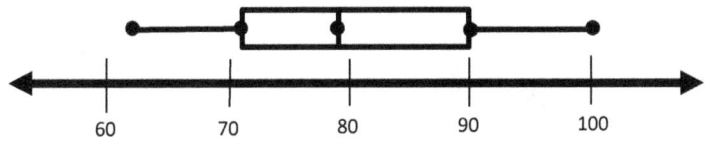

2) 11, 17, 22, 18, 23, 2, 3, 16, 21, 7, 8, 15, 5

Maximum: 23, Minimum: 2, Q_1: 6.5, Q_2: 15.5, Q_3: 19.5

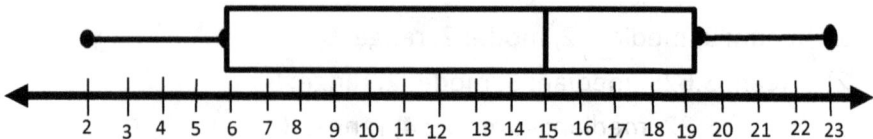

14–4 Bar Graph

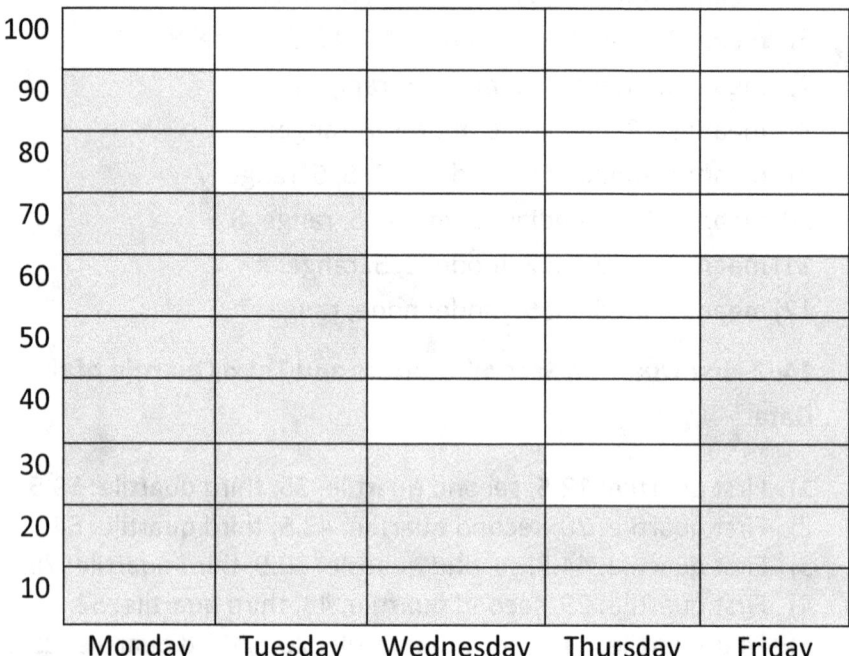

14–5 Stem–And–Leaf Plot

1)

Stem	leaf
7	2 4 9 9
8	3 6 8
9	1 5 7

key: 86

2)

Stem	leaf
1	9
2	6 6 6
3	3 7
4	8 8 9

key:

3)

Stem	leaf
4	1 2
5	3 4 4 8
6	5 5 7 9

key: 65

14–6 The Pie Graph or Circle Graph

1) red
2) 15%
3) black
4) 23%
5) 13%

14–7 Scatter Plots

X	Y
1	20
2	40
3	50
4	60

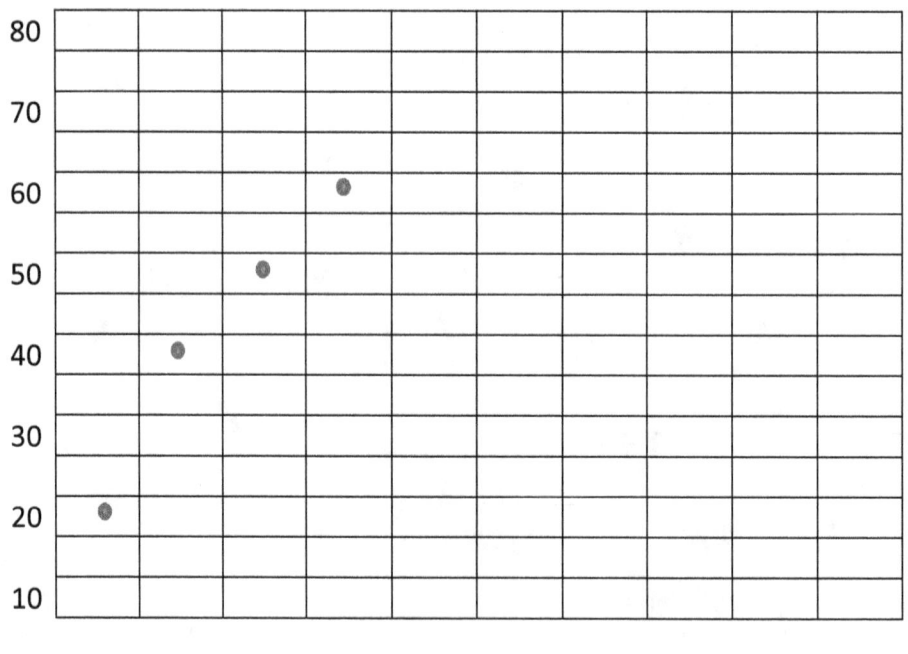

Arithmetic and Pre-Algebra Workbook

"Effortless Math" Publications

Effortless Math authors' team strives to prepare and publish the best quality Mathematics learning resources to make learning Math easier for all. We hope that our publications help you or your student learn Math in an effective way.

We all in Effortless Math wish you good luck and successful studies!

Effortless Math Authors

Online Math Lessons

Enjoy interactive Math lessons online

with the best Math teachers

Online Math learning that's effective, affordable, flexible, and fun

Learn Math wherever you want; when you want
Ultimate flexibility. You can now learn Math online, enjoy high quality engaging lessons no matter where in the world you are. It's affordable too.

Learn Math with one-on-one classes
We provide one-on-one Math tutoring online. We believe that one-to-one tutoring is the most effective way to learn Math.

Qualified Math tutors
Working with the best Math tutors in the world is the key to success! Our tutors give you the support and motivation you need to succeed with a personal touch.

Online Math Lessons

It's easy! Here's how it works.

1- Request a FREE introductory session.

2- Meet a Math tutor online via Skype.

3- Start Learning Math in Minutes.

Send Email to: info@EffortlessMath.com

Or Call: +1-469-230-3605